Childhood and Biopolitics

Studies in Childhood and Youth

Series Editors: **Allison James**, University of Sheffield, UK, and **Adrian James**, University of Sheffield, UK.

Titles include:

Kate Bacon
TWINS IN SOCIETY
Parents, Bodies, Space and Talk

David Buckingham and Vebjørg Tingstad (*editors*)
CHILDHOOD AND CONSUMER CULTURE

Tom Cockburn
RETHINKING CHILDREN'S CITIZENSHIP

Sam Frankel
CHILDREN, MORALITY AND SOCIETY

Allison James
SOCIALISING CHILDREN

Allison James, Anne Trine Kjørholt and Vebjørg Tingstad (*editors*)
CHILDREN, FOOD AND IDENTITY IN EVERYDAY LIFE

Nick Lee
CHILDHOOD AND BIOPOLITICS
Climate Change, Life Processes and Human Futures

Manfred Liebel, Karl Hanson, Iven Saadi and Wouter Vandenhole (*editors*)
CHILDREN'S RIGHTS FROM BELOW
Cross-Cultural Perspectives

Helen Stapleton
SURVIVING TEENAGE MOTHERHOOD
Myths and Realities

Afua Twum-Danso Imoh, Robert Ame
CHILDHOODS AT THE INTERSECTION OF THE LOCAL AND THE GLOBAL

Hanne Warming
PARTICIPATION, CITIZENSHIP AND TRUST IN CHILDREN'S LIVES

Rebekah Willett, Chris Richards, Jackie Marsh, Andrew Burn,
Julia C Bishop (*editors*)
CHILDREN, MEDIA AND PLAYGROUND CULTURES
Ethnographic Studies of School Playtimes

Studies in Childhood and Youth
Series Standing Order ISBN 978–0–230–21686–0 hardback
(*outside North America only*)

You can receive future titles in this series as they are published by placing a standing order. Please contact your bookseller or, in case of difficulty, write to us at the address below with your name and address, the title of the series and the ISBN quoted above.

Customer Services Department, Macmillan Distribution Ltd, Houndmills, Basingstoke, Hampshire RG21 6XS, England

---

Childhood and Biopolitics

Climate Change, Life Processes and Human Futures

Nick Lee
University of Warwick, UK

First published 2013 by
PALGRAVE MACMILLAN

Palgrave Macmillan in the UK is an imprint of Macmillan Publishers Limited, registered in England, company number 785998, of Houndmills, Basingstoke, Hampshire RG21 6XS.

Palgrave Macmillan in the US is a division of St Martin's Press LLC, 175 Fifth Avenue, New York, NY 10010.

Palgrave Macmillan is the global academic imprint of the above companies and has companies and representatives throughout the world.

Palgrave® and Macmillan® are registered trademarks in the United States, the United Kingdom, Europe and other countries.

ISBN 978–0–230–25227–1

This book is printed on paper suitable for recycling and made from fully managed and sustained forest sources. Logging, pulping and manufacturing processes are expected to conform to the environmental regulations of the country of origin.

A catalogue record for this book is available from the British Library.

A catalog record for this book is available from the Library of Congress.

To Casper, Johanna, Saskia, Liza and Cassidy

Contents

Acknowledgements

Thanks to Anna Sparrman and all at Tema Barn, Linkoping University, Sweden, for your time and your helpful comments on some of this material.

Johanna, thanks for your patience and support.

1
Children as Human Futures: Children as Life-forms

At least two relationships are regularly formed between children and the 'future'. From the perspective of parents and other family members, children often represent, among other things, the survival of highly valued characteristics, attributes and traditions beyond individuals' death. From the perspective of the political and geographical entities know as 'states', children represent an opportunity to shape and to secure a future for the populations they govern. By intervening in the health or education of a child, they can hope, for example, to increase the likelihood that, as an adult, she will be a net contributor to state finances.

In these two senses then, children are often understood to be special kinds of human – 'human futures' – at once bridging the gap between the present and the future and being the material from which the future will be made. Of course, making connections with futures is not the only form of 'time-work' that societies use their young for. As I have already observed, children are sometimes taken to embody characteristics of previous generations. This book however is particularly concerned with the place of children as resources in forms of time-work that are conducted by individuals (including children themselves), states and other organisations that are intended to shape the future from the stance of the present. If 'shaping futures by using people as resources' sounds 'political', then I am setting the right tone. As I will argue later, such forms of time-work are part of the basic architecture of states. For the most part, the fact that this work takes place at all is buried and obscured in the calmly shared assumptions that compose the official adult world and official versions of childhood. Where these assumptions are breached or challenged however by social or technological change and time-work is brought to the surface, hot controversy can follow. This is how childhood can be at once a matter of consensus and of conflict.

The study of childhood by social scientists has long been based on these relationships between children and futures. Throughout the twentieth century, for example, concern for children's proper development and socialisation was inspired by their status as bearers of the future. The variously coercive and disciplinary role that child science, in consequence, played in childrens' lives has been thoroughly examined (Burman 2007). I suggest however that, today, two trends that have very broad implications are changing our relationship both with the future and with children. First, due in part to the issue of climate change (IPCC 2007), the future itself is beginning to look ever less open to and accommodating of human plans. For decades, economic growth, based largely on the use of fossil fuels, has been the goal of governments and has become a measure of good governance around the world (Mahon and McBride 2009). Children of the 'developed' or 'minority world' countries in particular have, for 50 years or so, been raised in accordance with a pair of assumptions about the future:

- Year on year, natural resources will be used in greater quantity
- Each generation will see greater wealth and well-being than the last.

Today, however, fossil fuel stocks are running low and their costs, both in terms of purchase price and in terms of the expected climate change their use brings, are rising. A similar story can be told about many other resources such as water and agricultural land, where climate change, increasing population densities and economic forces are working together to create shortages. It often seems that our usual means of securing the future are failing and, in some circumstances, are producing new insecurities. If this is the case, then childhoods, considered as a key element of these security strategies, are likely to be affected. A large part of this book examines such security strategies and explores the implications of their functioning and of their failure for children's place in global bio-politics.

Second, for many years, a certain human quality of 'plasticity' (Giddens 1992) was understood to be concentrated in the first two decades of the life-course. The connection between this plasticity and the futurity of the young was proverbial: As the twig is bent, so the tree shall grow (Stainton Rogers and Stainton Rogers 1992). Today, developments in the bio-sciences appear to be creating new plasticities and redistributing them throughout the life-course. In some areas, this appears to increase choice about key life events. Pregnancy and life-span, for example, are ever more open to deliberate influence. Further, pharmaceuticals are in development that could increase adults' and children's capacities to learn (Office of Science and Technology 2005). Perhaps most fundamentally, techniques are becoming available that can extend 'plasticity' to the genetic level, extending the ability to shape the young to points well before their conception. A good deal of academic coverage of these issues currently takes the form of speculation about the ethical quandaries that might face individuals or policy makers if some, as yet impossible or unproven, technology became available. Examples here include Harris (2007) on 'smart drugs' and Prout (2005) on human cloning. Explorations of this kind are clearly of value, revealing as they do, ethical commitments that may otherwise go unexamined. I will pay them due attention.

Consistent with my focus on the bio-politics of childhood, however, as I address plasticities, I will have two biases. Rather than focus in great detail on ethical decisions about plasticities, I will tend to place plasticities in a global bio-political context. Thus, from my point of view, the most interesting thing about 'smart drugs' in education is not whether they should be banned, regulated or given out for free, but the specific contexts of personal investment and time-work in which taking a pill to make oneself 'smarter' may seem a sensible thing to do. Second, many of the plasticities of significance for childhoods do not depend on future technological innovations. When considering such developments, I tend to the view that the future is already here but that it is unevenly distributed. Thus, the recent collaboration of the Gates Foundation and the pharmaceutical company GlaxoSmithKline to make vaccines available to children in developing countries at relatively low cost (see Chapter 5) interests me more than human cloning.

Together, climate change and developments in the life sciences pose a wide range of challenges and opportunities for individuals, families and states. They affect our relationships with the future through the medium of our existence as biological creatures – as eaters of food and drinkers of water who are composed of cells and organs. In this sense a good deal of today's politics of childhood is 'bio-politics'. Much of this book is taken up with defining those challenges and examining existing and emergent responses. But, in my view, understanding and responding to these developments poses particular challenges for childhood studies as a field. Over the past decade, a 'new paradigm' of social constructionist research (James and Prout 1997) has been very influential. Broadly speaking, this approach has sought to challenge the image of children as human futures. It is based on the view that identifying children with the future has tended unjustly to silence children as present-day members of society and to sanction their use as a resource instead of their inclusion as citizens with views and preferences of their own. The central critical claim of the 'new paradigm' is that since childrens' future as adults has been so important for states, it has tended to outweigh childrens' present-day lives, experiences and points of view. Thus 'new paradigm' work tends to emphasise childrens' 'agency' and childrens' 'voice', insisting, against what it takes to be dominant arrangements and perceptions, that children can act, make decisions and form opinions about their lives and other topics in independence from adults.

For good strategic reasons however, of which more later, this 'new paradigm' has tended to downplay children's existence as biological creatures. I am going to argue that if the two trends of climate change and life science development are together bringing the 'biology' back to human futures, and if childhood studies is to help individuals, families and states with the challenges they face, then ways must be found to make the social scientific study of childhood comfortable with seeing children as life-forms. But first I'd like to clarify a few points about childhood and 'the future' as they often appear when seen through the lens of today's fears and hopes.

Children, futures and survival

For many adults, children are an embodiment of hope and of consolation. A child can function as an anchor of meaning in a world of rapid change and unexpected loss. For some parents, having a child can offer consolation for the fact of their mortality. When a child is seen as embodying hope for a better future, this can offer some compensation for adults' struggles in the present, along with the motivation to keep going. Children also present grounds for the hope that something of ones' self will last beyond the course of one's own life. For some, it is the family name and reputation that is taken to survive with the child, for others it is their values, or tenets of their faith. The idea of survival can be projected onto the inheritance of physical or behavioural characteristics – 'he has his father's eyes' – and even onto the portion of genetic 'make-up' a child owes to a biological parent.

A view of children as hope and consolation can also be shared by adults who are not themselves parents. Even childless atheists may still have enough 'biophilia' (Kellert and Wilson 1995) – love of living things – to see and to value children as highly adaptive and crafty participants in the experiment called 'life', just like themselves. These hopeful associations help to motivate a good deal of adult investment in children. It can also make adults quite insistent with children about their beliefs and behaviour, about 'right' and 'wrong'. Life can be uncomfortable for children who do not live up to the hopes invested in them or whose existence fails to console the adults around them. Grouping them together, we can call these various associations of children with hope and with consolation the 'survival fantasies' of adults. In all likelihood these fantasies have a long, and mostly unwritten, history.

In recent years however, in debates on climate change and bioscience, positions have emerged which, if they are accurate, would seem either to destroy these survival fantasies or simply to make them irrelevant. We have, for example, James Lovelock's (2009) forecast of near-future, dramatic and long-lasting global temperature rise, which would place the survival of the human species (among many others) in question. He argues that global warming (DiMento and Doughman 2007) has already gone beyond the point at which it could be halted or reversed by changes in policy and that the best we can do is prepare for an inevitable decline of human population and fortunes. Lovelock originally gave us the idea of the Earth, its geology, seas, ice caps and life-forms as one self-regulating system which he called 'Gaia' (Lovelock 1979), the ancient Greek name for the mother goddess of the Earth. For many years his 'Gaia' hypothesis was not taken at all seriously by the scientific community. The facts of global warming have changed that. He now suggests the we, humans, or rather a specific form of human life that is based on burning fossil fuels like oil, coal and gas, are today being 'regulated' out of existence as the Earth's interdependent systems try to adapt to the rapid rise in carbon dioxide (CO_2) concentrations in the atmosphere caused by our daily activities. So there is a clear prospect of death, of an end to children and of an end to survival that is advancing on us and moving quite rapidly.

On the other hand, we have the gerontologist Aubrey de Grey (De Grey and Rae 2008). He argues that scientific progress in the understanding and prevention of ageing is such a strong trend that, with the right strategy and resources, individuals born today could live for 1,000 years. The trick is to use whatever available rejuvenation techniques one can to stay alive long enough for the next bio-medical leap forward, and then to catch that wave soon enough to make it to the next. For now, restrict your calorie intake, take exercise, don't smoke or drink alcohol. Then, when you are in your 70s, stem cell research (Panno 2006) or nano-technology (Jones 2007) will provide something to keep you going till 100. By then, a further set of as yet unimagined techniques will be available. De Grey envisages a 1,000-year life rather than an eternal life because of the statistics of fatal accidents. Live long enough and a traffic accident will probably carry you off. Nevertheless, if there is no death, at least not for absolutely ages, there would be need for children and the second-hand survival that they offer.

Lovelock (2009) does not rule out the possibility that some humans may survive. If the right steps are taken, a percentage of the existing population may cling on in climate change 'lifeboats', such as Northern Europe, where a temperate climate persists. Similarly De Grey and Rae (2008) do not suggest the end of death entirely. But these visions trade in the same terms (death, life, continuity, termination) as do survival fantasies. They come to us from the brightly lit realm of scientific debate, disturbing our dreams of continuity and harmony between the generations. If there is no future, how can children 'be' the future? If the future involves 500-year-old people, or indeed cloned humans, will there be space for or need for children as we currently understand them? These visions share an emphasis on the fact that humans are biological creatures and that we are in a relationship of mutual dependency with other life-forms. Further, today we are such creatures that we can intervene in life processes at both the global and molecular levels. Thus, one commentator even argues that we now can and should 'enhance' our own evolution (Harris 2007) designing ourselves and future generations as we see fit.

The question of what can and should be made of children is hardly new (Lee 2001), but together, climate change and the life sciences raise the stakes and increase the latitude of that question. This book explores and illustrates a few of the issues that lie between the speculative extremes defined by Lovelock and De Grey. It addresses young people's cognitive powers, vaccines to prolong life and health and climate change. Further, it develops the tools to understand these issues so that childhood studies is prepared to play its part in shaping human futures.

What does 'bio-politics' mean?

So far, I have suggested that a good deal of thought and feeling about childhood is structured by what I have called adult 'survival fantasies'. I have further suggested that these fantasies, though important, are not adequate to the task of understanding childhood in a world where climate change and bio-science foreground our existence as biological creatures. In one sense, then, real-life bio-politics are the issues of climate change and sustainability, medicine and well-being that lie between the extremes marked out by Lovelock and De Grey. A set of ideas is available, however, that is designed precisely to understand the social and political significance of our existence as bodily human creatures. Chapter 2 sets these ideas out with greater clarity and explores some of the tensions between them, but for now I need further to explain the term 'bio-politics' that I have chosen to pair with 'childhood' in my title.

The historian and political philosopher Michel Foucault (2007) coined the term 'bio-politics' in a series of lectures he gave in the late 1970s. For him, bio-politics is the special field of politics, of argument, strategy and bids for control (successful or otherwise), which became available in the eighteenth century with the realisation that, apart from anything else, humans are also a 'species'. For humans to be a species means that they are biological creatures like any other, with their own needs, abilities and behavioural tendencies. This may seem obvious to us now, but Foucault (2007) carefully shows that it has not always been so. For many years, after all, the primary European frame for understanding humanity concerned God, sin and redemption. He describes the emergence among the governments and policy makers of European states of a bio-political way of thinking about and governing populations through the eighteenth century. The bio-political relation between governments and populations has since developed and spread till today and it is taken for granted by many in power. Let me sketch out Foucault's account of the bio-political.

One thing rulers have long done, and still do today, is say what the ruled *must not* do, by defining what is against the law. Foucault calls this form of power 'juridical'. The definition of some forms of killing as murder and the provision of punishments for murders is a good example here. Another thing rulers can do is say what the ruled *must* do, and exactly how and when they should do it if they want to become 'good' people. This is the form of power that Foucault calls 'discipline'. It is still widespread today. For example, if unemployed, a person in the United Kingdom must take steps to show authorities that they are actively seeking work and prove that they are taking the right steps at the right time so that they qualify for state assistance. Alternatively, you might think of a doctor giving a child or parent advice on healthy eating, or the gym instructor designing an exercise programme for a client.

For Foucault, 'bio-power' is quite distinct from juridical and disciplinary forms of power, even though it may use these other forms of power to reach its ends. Rather than starting with an idea of what is wrong or right for individuals to do, bio-power starts with a goal, say the reduction of rates of a certain disease amongst a certain group of people, and then tries to understand how the needs, abilities, proclivities and susceptibilities of a population may be harnessed to reach this goal. For Foucault (2008), by the late twentieth century this field of 'bio-politics' had come to focus on the goal of providing populations with economic security and had helped create strong links between the practice of government and particular economic theories forging 'neo-liberal' forms of rule. We will return to that later in the book. For the moment, however, we need to note that others inspired by Foucault have expanded on this concept of 'bio-politics' to include issues of the framing of relationships between life processes and lifestyles (Agamben 1998; Rose 2007). These issues, covered in greater detail in Chapter 2, become quite important as my arguments develop across the book.

A new environment for childhood studies

Together Foucault and those he has inspired offer us clues about how students of childhood might respond to the new environment that childhood studies finds itself in. This environment, as my introduction suggested, is defined by two major tendencies. The first tendency is the increasing recognition that human societies, lifestyles and identities are inseparably bound to the wider life of the planet. If we exclude the colonisation of other planets as at present impractical, we have no choice in the fundamental matter of this relationship. Even though the wealthier among us are still able to afford lifestyles that seem separate from wider flows of resource, in small eddies, as it were, of plentiful water and abundant oil, it is now clear that this separability (Lee 2005) cannot last forever and that it maximises costs to the poor. The second tendency is the growing desire that both shapes and is shaped by bio-science, to intervene in human life processes so as to prevent disease, prolong life or enhance our capabilities. The achievements and promise of such fields as genetic medicine (Wright and Hastie 2007) and neuroscience (Blakemore and Frith 2005) are considerable and exciting. With one or two exceptions (Prout 1999, 2005) childhood studies has so far given very little attention to these issues. In the next section, I will offer an account of why that is. It will also become clear why I think this new bio-political environment requires that childhood studies adjust some of its guiding assumptions and broaden its range of concerns.

Throughout this book, I will examine the events and relationships that are forming today's childhoods and shaping children's lives as the consequences of our two key trends develop. But I will also be searching for and developing new ways of thinking that can help today's students of childhood make a contribution to human futures. The first step on the journey is to clarify the challenge that exists for students of childhood, as I see it, if they wish to engage with climate change and bio-science.

Childhood studies: The story so far

Since its formation in the late twentieth century, the social scientific field of childhood studies has given relatively little attention to the fact that children are biological creatures. It has instead emphasised the social aspects of childhood covering the historical, social and cultural diversity of childhoods (Corsaro 2004; James and Prout 1997), children's relationship to democratic participation, rights and voice (Alderson 2000) and their levels of social and cultural competence (Hutchby and Moran Ellis 1998) among other topics. The reasons for this apparent oversight lie in the arguments and struggles that initially gave rise to childhood studies. As I will suggest, these arguments of the late twentieth century were about the degree to which childhood can and should be standardised. In what follows I will further suggest that, for strategic reasons connected with debates over a standard childhood, childhood studies of the last two decades has tended to treat the 'natural', the 'biological' and the creaturely aspects of childhood as its enemy. I will then argue that conditions have changed to such a degree that this strategy and many of the assumptions that have built up around it must now also change.

Childhood, considered as a period of social, moral, intellectual and practical preparation, has long been of central importance to industrialised societies, among others (Donzelot 1979; Lee 2001). The twentieth century saw a Western childhood globalised as a set of aspirations and crystallised in the worldwide expansion of mass education systems (Zajda 2005). Across the century, a chain of associations was embedded in the cultures and institutions of many nations that linked the young with hope for better futures. This gave childhood a peculiar temporality – each child a future in formation, a human future – and a special place in strategies for economic growth and competition worldwide, making children a site of investment and a practical way of influencing the future from the standpoint of the present.

Across the twentieth century, children became valued as a human resource of a peculiarly malleable kind, especially receptive to education and socialisation, such that child labour could widely come to be seen as relatively wasteful of resources (Basu 1999). Wherever national progress and well-being were defined in terms of economic growth, for example a rising Gross Domestic Product figure, children were largely, but not exclusively, understood as human fragments of the future available in the present for shaping and design towards the end of national wealth creation.

Thus, childhood, defined as a phase of economic dependence, personal development and educational investment that was unique in the human life-course, emerged as a global standard form of life for children and the key concept in the government of humans as life-forms. In 1989 the United Nations Convention on the Rights of the Child (UNCRC; www.unicef.org) gave explicit expression to these temporal, developmental and economic relations in those of its articles related to the provision of educational and developmental resources to children. It expresses this standard in the form of 'rights'. Article 27, for example, states,

1. States Parties recognize the right of every child to a standard of living adequate for the child's physical, mental, spiritual, moral and social development.

(www.unicef.org)

And Article 28,

1. States Parties recognize the right of the child to education, and with a view to achieving this right progressively and on the basis of equal opportunity, they shall, in particular:

(a) Make primary education compulsory and available free to all...

(www.unicef.org)

Not all children, of course, have access to such standard provision despite the UNCRC. Many countries of the majority world lack the necessary resources, for one reason or another. Further, some of the children who do have access to the standard model of childhood nevertheless find themselves silenced by the many positions of dependence on adults that it offers them.

In response to these developments, the social scientists and humanists who formed the field of childhood studies in the late twentieth century focussed their work on the limitations of this standard view of childhood. They argued that standard childhood had the potential to reflect and strengthen relationships of power between children and adults and argued that these power relations were sometimes unjust. This agenda has drawn attention to the dangers to those children whose life circumstances mean that they fall short of the standard as well as the discontents of those defined by it.

If childhood were simply a natural phenomenon and thus fundamentally the same for every single child, then a single standard view of childhood built around 'nature' should work well for every child. In order to highlight those instances where the standard view did not work well, childhood studies argued that childhood is not a natural phenomenon, that it is not universal and that, therefore, no single standard will work well for every child. This is one reason why so much attention has been devoted to social and cultural diversity of childhoods and children's relationship to democracy, rights and voice over the years. Thus, it has been argued that rather than there being one childhood that was simply a 'natural' stage of human growth, there were in fact many different childhoods created by different societies and cultures to serve purposes of their own. There were even some circumstances where the term 'childhood' had no meaning at all (Ariès 1962). Casting childhood as a social phenomenon rather than a natural one had a key strategic advantage. Anything social or cultural could, in principle, be changed. When authors in this tradition say that childhood is 'socially constructed' (James and Prout 1997), this diversity and possibility of change is what they mean.

Arguably, this research programme has successfully drawn the attention of policy makers, as well as other academics, to childhood as a socially constructed 'form of life' (Rose 2007). But there are now good reasons to think that we need to supplement this programme with a new focus on children as 'life-forms' (Rose 2007). Instead of simply rejecting the view that childhood is a natural phenomenon, we now need to take 'biology' or 'life processes' into account. One important feature of this is the recognition that there is no single univocal 'nature'. In Europe many years ago, when 'natural law' was another word for God (Foucault 2007: 233), to say that a feature of human society was 'natural' was to place it beyond critique and to make it very hard to imagine it ever changing. If the word 'biology' had the same implications today as 'nature' used to – denoting the unchangeable and the universal – then taking it into account would just mean losing the ability to criticise children's place in society along with the ability to imagine other arrangements. Fortunately that is simply not the case. As I will argue in the following section, discussion of the 'biological' and the 'social' and of the relations between them has moved on quite rapidly in recent years.

Natural diversities, inflexible cultures?

In the 'social constructionist' programme of childhood studies described above, the important things to pay attention to were norms, values, power relations, discourses – all things made by, between or with the involvement of human beings, things that many adults would feel they should have 'a say in' or a degree of influence over. The democratic impetus for a good deal of childhood studies was to build a sense that children should also have a degree of influence over those same things so that the limitations of the standard view of childhood could be recognised. Unfortunately, this successful strategy carries with it a set of assumptions that nowadays in the context of climate science and the life sciences more broadly look quite odd. Simply stated, it portrays the 'natural' as old, homogeneous and slow to change and the 'social' as young, diverse and fast changing. It supposes that the 'social', the 'cultural', the sphere of human relations are the location and source of diversity and change, and that the 'natural', the 'biological' or the creaturely aspects of human life are the location and source of the universal and unchanging. These associations may once have seemed necessary to the task of challenging the standard view of childhood. But do they withstand scrutiny today?

In the two decades since the publication of *Constructing and Reconstructing Childhood* (James and Prout 1997), evidence for both 'natural' diversity and 'cultural' universalism and inflexibility has grown that would seem quite counterintuitive from within the 'social constructionist' programme. The Human Genome Project (US Department of Energy 2009) was completed in 2003. It identified each of the tens of thousands of genes in human DNA and determined the sequences of the three billion chemical 'base pairs' that make up that DNA. These figures alone, hinting at the sheer complexity of life processes, the proximity to chaos that lends them such sensitivity and adaptability (Lewin 1994), make any view of 'nature' as a dependable, unarguable, changeless ground look rather flimsy.

As well as giving us these details about human similarities, however, the human genome project also paved the way for new investigative and technological responses to individual variation and human diversity. One result is the research area called pharmaco-genetics that seeks to tailor drug treatments to individuals' genetic characteristics. Another is an increased sense of precision in the notion of individual susceptibility to disease and other 'dysfunction' (Rose 2007). To re-emphasise biology in today's context is in no way to assert that childhood is 'universal' or that it can or should be standardised.

Should the 'social' or the 'cultural' be considered innately diverse and changeable? The Kyoto Protocol (www.unfcc.int/kyoto) was pro-duced in 1992 and came into force in 2005 establishing legally binding commitments for the reduction of emissions of four 'green-house' gases associated with climate change, including CO_2. If our culture was composed solely of discourses and values, we could prob-ably have met its targets quite easily by talking down our emissions. Talk is important, but culture is about materials and material relations too (Latour 2007). Changing a 'culture' means, among other things, decommissioning coal-fired power stations, re-drawing road systems to allow safe cycle travel and finding economically viable alternative energy sources. It is an awkward and time-consuming process. The Kyoto Protocol, for instance, had to find a way to fit the ambition of reducing emissions into the already established, extremely elab-orate relations of trade and manufacture that are inseparable from industrial lifestyles and cultures. The result was an emissions 'trad-ing system' which, at best, has had a questionable impact (Giddens 2009). When we consider this globalised way of life that is so slow to change, I would suggest that the association of the social with diversity and change does not seem justified.

We are now acutely aware that, together, our bodies, our cultures and our technologies have inputs and outputs of energy. Whether we like it or not, whether we intend it or not and whether or not we are aware of it, they give pattern to the movement and distribution of molecules, altering, on the largest scale, atmospheric CO_2 concen-trations, and on the smallest, interacting with, say, in the case of some pharmaceuticals, cholinesterase inhibitors in our brains. Given this, diversities are to be found everywhere and speeds of change are always multiple. We can no longer look to nature as the unchang-ing ground of our activities (Serres 1995) or to culture as liberation from materiality (Latour 1993). Given time, everything is changeable in this one world. In my view, unless childhood researchers take this into account they will not be able to keep pace with our rapidly changing world or be able to chart the many temporalities of new and emerging forms of time-work and bio-politics. The problem is that it is not entirely clear what can replace the 'binary frame' that opposes an old, slow 'natural' sphere, to a young, fast 'social' sphere. Over the next two chapters, it will become clear that I am far from alone in problematising binary frames around childhood and in offering alternative approaches. What I hope to do is draw on this debate to create a 'bespoke' framework, tailored to the needs of childhood studies as it encounters contemporary bio-politics. This will involve the description of three new conceptual figures: 'multiplicities of childhood' that I describe in Chapter 2 and 'biosocial events' and 'biosocial imaginations' which I describe in Chapter 3.

Challenge, change and strategy

Research fields either respond productively to changes in the world around them or fade into irrelevance. Over the past 100 years or so, research on childhood has adapted rather well. Childhood literature is studded with examples of how, in the past, the core questions and methods characteristic of various psychological and sociological studies of childhood have been developed in intimate relation with changing policy agendas, political debates and extant economic structures. This adaptation does not imply that the study of childhood has passively reflected its environment or that childhood research lacks a degree of autonomy. Rather it suggests that the manner in which problems and responses are formulated within the field has always had a strategic aspect. Childhood researchers have frequently adapted their concerns, techniques and sensibilities to new conditions, seeking the best angles and opportunities to influence outcomes. Sometimes all the adaptation that is required is a further elaboration of an existing set of questions and methods. Occasionally something bolder is called for. In my view, recent developments in the life sciences and around climate change make this one of those occasions. I'll say more on this shortly but first I'd like to sketch the last major adaptation in the field and say something about where it has left us.

In the early to mid-twentieth century, psychologists and sociologists of childhood were the sort of people you would talk to if you wanted to know what environments are best for children's development or why and how children should be socialised. Today's field of childhood studies, with its lawyers, historians, cultural critics, psychologists and sociologists, still addresses these concerns, but it covers much more ground. Over the last few decades, the agenda for childhood studies has been significantly shaped by economic, cultural and regulatory forms of 'globalisation' (Prout 2005). From the mid-twentieth century on, many matters of trade, cultural production and legal and quasi-legal regulation became 'globalised'. Container shipping and information technologies built new interdependencies between distant global regions. New media forms like satellite television and the Internet made fresh links between cultures. Projects aiming at the international regulation of 'rights' were formulated and enacted, and organisations like the International Monetary Fund and World Bank tried to produce a global consensus on economic policy to integrate minority and majority world economies. Each of these developments contributed to enormous changes in people's daily and intimate lives that have been experienced around the globe. 'Traditional' family relations have been challenged, urban populations have outgrown rural ones and the viability of small-scale agriculture placed in question. At the same time, wealthier populations worldwide have grown used to defining themselves in terms of their consumption patterns and have significantly raised their expectations of economic growth and for their children's futures.

Such globalisation provided the context in which a pair of distinctively late-twentieth-century questions about the nature of childhood and children emerged to shape the agenda of childhood studies:

- Is there a 'universal' childhood fundamentally the same across cultures?
- To what extent are children capable of independent thought and action?

The first question arose out of the tension between the manifest diversity of global childhoods and the ambition globally to regulate those childhoods through a universal childrens' rights instruments – the UNCRC. As I've suggested above, proponents of childhood diversity often worked against the idea that childhood is a fundamentally biological phenomenon, preferring to view childhood as 'socially constructed'. Some criticised versions of 'universalist developmentalism' that depicted all childhoods as involving fundamentally the same processes of growth and learning. Positions in favour of there being one universal or basic 'childhood' underlying all diversity were often taken in attempts to legitimate universal children's rights. Smith (2010) covers this ground in depth. The second question of how much independence children should have and can cope with arose in response to the de-traditionalisation of family life and the emergence of consumer societies that, respectively, tended to challenge traditional patriarchal decision making and to encourage people to see their lives in terms of making independent choices (Lee 2001).

The social changes brought about by globalisation were radical but, as a field, childhood studies was able productively to respond to them. It did so by turning its core commitments into questions. In the early twentieth century, children were largely understood quite simply as 'becomings', lesser than and less capable than adults (Lee 2005). Childhood research of the time investigated the best ways to bring them from this state into one of fully functioning adult 'being' through psychological development and socialisation. But by the late twentieth century, one major issue was whether children were best understood as 'becomings' at all and whether they could or should instead be considered 'beings' in parity with adults.

It may not always be apparent, but this adaptation has drawn childhood studies into a close relationship with a question that lies at the heart of Western social and political thought: What is the relation of humans to nature? I will return to the roots and implications of this question in greater detail in later chapters, but for the moment, I'll make just a couple of observations. First, the question of the human/nature relation, as posed above, cannot simply be 'answered'. This is because it houses an ambiguity that is constitutive of many Western philosophical and cultural traditions in which people are considered both as animals and as something more. This ambiguity sets up tensions around people's dependence on and ability to distance themselves from life processes (Elias 1994). Thinking about my personal relationship to 'nature' it is clear that in many ways I 'transcend' it and thereby gain autonomy from it. For example, I use natural products, like wool, to protect my body from the cold. I also take part in political decision making that, at a distance, controls how natural resources are used. So in some ways, I am separate from 'nature' and able to act on it from a distance, as it were. At the same time, however, I can only do this because I am made up of 'nature'. Without a brain or a heart, I would have great difficulty voting or choosing a t-shirt. In this sense I am both independent of and dependent upon nature. I've argued elsewhere (Lee 2001) that the being and becoming question raised around childhood is, like its close neighbour the 'human nature' question, an irresolvable one. It is a paradox, in the sense that it contains equally valid but contradictory commitments. Adults and children alike can reasonably be considered independent beings or emergent dependent becomings or both, depending on circumstances (Lee 2005). But there is much more to say.

How many 'natures'? How many categories?

I've suggested so far that the 'being/becoming' tension around childhood is a version of that ancient question of the relation between humanity and nature and that, as such, it cannot be resolved. As I will argue in greater detail in Chapter 2, this question is part of a frame that has been used for understanding human existence for a couple of thousand years. Of course just because a question cannot be resolved does not mean it is not productive. However, there are reasons to think the question of the relation between humans and nature is often poorly stated. It tends to treat 'nature' as if 'nature' were all fundamentally the same thing, as if the life processes that go on in my brain were examples of the same 'nature' as the life processes that allow sheep to grow wool. 'Woolly thinking' aside, it is pretty clear that my relation as a human to my own cognitive processes is rather different from my relationship as a human to the processes that create wool.

In my view, this picture of nature as single thing is a trap. It certainly helps us formulate certain sorts of questions about childhood:

Are children beings or becomings?
Are people shaped by nature or nurture?
Is childhood natural or cultural?

But in doing so it requires us to commit to the idea that just two categories are enough to describe, let alone generate questions about, the great diversity of childhoods and children. This sort of commitment is often referred to as 'dualism'. I'll use that term interchangeably with the phrase 'binary frame'. A brief examination of commitments common across the field of childhood studies will emphasise the field's indebtedness to it:

- The key conceptual resource for understanding childhood is the cultural distinction between human being and human becoming.
- To be a 'human becoming' is to be, as yet, relatively distant from social norms and from cultural competence, to be on one's way to a more developed state, and in a sense to be closer to nature. Membership of this category confers relatively low social status.
- To be a 'human being' is to be self-determining and driven by social imperatives above biological ones. Membership of this category confers relatively high social status.
- The more childhood researchers can convert the image of children and constructions of childhood from 'becoming' to 'being', the more ethically adequate and politically progressive our research and arguments will be.

I don't find the binary frame particularly credible. As Deleuze (1991) has it, two-category schemes hang on the world 'like baggy clothes', making contact here and there, but mostly just pursuing their own line. It is important to recognise those few places where a binary frame looks like a good match for a state of affairs. But it will be just as important for my argument to find ways to avoid over-application of a binary frame so as to make space for new habits of thought and conceptual methodologies.

Up until quite recently, the problems with the binary frame haven't mattered much. As I've stated, childhood studies has been able productively to respond to the enormous changes brought by globalisation and has done so by expanding its list of theoretical and empirical research questions. It has had plenty of work to do along the way and has found the two-category schemes like 'being/becoming' and 'natural/social' quite useful in this. But, as I will argue through this chapter, at a time of rapid advances in bio-science and heightened awareness of climate change, the limitations of this understanding of nature as singular and of the binary frame that underlies being/becoming, nature/nurture and nature/culture distinctions are ever more apparent. In my view, their value as a basis of enquiry relevant to the contemporary world is fast fading away. This is because the more that is known about life processes, about how much we depend on them and about how sensitive they are to the social processes we take part in the less obviously distinct 'nature' and 'social and cultural' life seem from each other. I find two examples particularly helpful here.

First, there is the relationship between the emergence of industrial societies and levels of atmospheric CO_2. Throughout human history, people have found ways to use natural resources like wood, coal and oil for power. Burning these resources means combining the carbon they contain with oxygen from the air. This releases energy and CO_2. The rate at which this took place was, for many years, so low that any effects it had on the atmospheric concentration of CO_2 was indistinguishable from variations due to other CO_2 sources. But then, as industrial manufacturing processes spread through Europe and North America from the mid-eighteenth century onwards, at first burning coal for steam and then oil for power and petrol for transport, the rate increased dramatically. The vast majority of climate scientists (IPCC 2007) draw a connection between the emergence of industrial society and the changing content of the atmosphere. If this view is correct then it is clear that industrialisation cannot be fully understood as simply a social, economic or political process. It takes place on a comparable scale to the global production of oxygen by plant life through photosynthesis, and the nature of its effects – changing the composition of the atmosphere – are comparable too. The binary frame that sets out to understand the world by applying a distinction between, say, nature and society is likely to get confused here because, as is now apparent, 'industrialisation' does not wholly fit into either category.

Second, there is growing evidence of biological and social processes working hand in hand. We might be used to thinking of biological differences between human populations as the sort of thing only racists take seriously. Alternatively we might think that biology changes so slowly compared to culture that we can safely ignore its effects, or even think of it as a stabilising influence, setting limits to cultural possibility. Recent research (Buroker et al. 2010), however, has compared the genomes of Tibetan and Han Chinese populations and found certain differences. A large proportion of Tibetans have genetic characteristics that enable them to live comfortably at the high altitudes and consequent low atmospheric oxygen levels common in their country. A rather lower proportion of Han Chinese have these characteristics. It seems that these differences date from about 2,700 years ago. This is the point in history at which a major migration from relatively low-lying land of modern-day China to the mountainous terrain of modern-day Tibet took place, dividing Tibetan from Han Chinese populations. In those thousands of years, processes of natural selection have increased the proportion of Tibetans whose genome allows them to cope well with low oxygen levels while no such selection has been experienced among Han Chinese. One result of this is that in a contemporary wave of migration of Han Chinese to Tibet, following the lines of new transport infrastructure, recent immigrants are suffering lower birth rates and higher infant mortality than their Tibetan neighbours.

It should be clear from these two examples that reference to 'nature' in general is simplistic and unhelpful. Instead of one 'nature' there are many life processes such as plants' photosynthetic production of oxygen, the CO_2 capture that tiny ocean organisms perform as they grow their calcium carbonate skeletons and the sources of adaptability and variation in human biological systems. Each of these has its own set of relationships with processes that we might normally think of as 'social'. Further the quality of these relationships can change in what I call 'biosocial events' (see Chapter 3). In Tibet, a specific capacity for change present in human biology emerged only as a population migrated. During industrialisation a new relationship formed between human lifestyles and the many life processes that affect relative concentrations of atmospheric gases. If, as I argue, childhood studies has been dependent on a binary frame and has not questioned its own tacit reliance on a singular nature, and if it is to make a serious contribution to the emerging world, then it will need to change.

What is this book for?

In this chapter so far I've made some bold claims and, I hope, generated a sense of urgency and expectation. As a reader, you might now be getting ready for me to bombard you with new rules and instructions on how to think. If so, relax. I'm hoping to do something rather more interesting for us both.

So far I've sketched out the reasons why I think childhood studies needs to change. Like a lot of other fields in the humanities and social sciences, it has been slow to recognise life processes and climate change and persistent in its commitment to two-category frameworks. Critique of binary frames has been well-rehearsed elsewhere (Haraway 1991; Latour 1993). But, positively, in the rest of the book I am going to try to build a framework for childhood studies that has a good fit with contemporary social concerns, policy agendas, political debates and economic structures. I am not driven in this by some desire to conform to the constraints of a new 'reality'. I am instead driven by the desire to develop the conceptual tools that lend critical and creative purchase in current conditions. To help me get there I will be drawing on the challenging ideas of a range of authors who have wrestled with similar issues to mine in the past. But in my view, it is no use looking to them as sources of authority as if they had all the answers, or 'following' them as if they knew the way. Taking a look at some of the periods of time I have mentioned in my argument so far (early and late twentieth centuries, 300 years, 2,700 years), it may already be apparent to you that I have a tendency to take the 'long view'. It should become clear in what follows that just as I try to make best use of resources from beyond childhood studies, I also intend to recognise and conserve existing strengths and sensitivities that have been developed in the study of childhood over the last century.

Conclusion: Childhood and human futures

So far I have argued that childhood studies needs to adapt to a new environment if it is to make a contribution to understanding and shaping the emerging future. Beyond asserting the growing significance of climate change and bio-science, however, I have said very little about what that future might be like or why I think it is important that a critical view of children's place in society should survive. They say it is very difficult to make accurate predictions, especially about the future. Nevertheless, it is worth sketching out some possible futures that are based on a few assumptions about the future availability of key resources and conceivable responses on the part of policy makers. I have drawn these assumptions from recent future-scoping projects (DCDC 2007; Glenn and Gordon 2007). This will yield some key issues that are likely to remain with us and make critical views of childhood just as important as ever.

Today, visions of the near future are beset by issues of climate change, the scarcity of natural resources such as oil and water and the unpredictable consequences of economic globalisation and the international interdependence it brings. At the time of writing, the world is suspended between a recent global economic recession and the end of the age of oil, and policy makers and communities are struggling to respond effectively to climate change. For good measure, add in the ageing of the global population that is turning youth into a scarce resource. Until recently, limits to global economic growth appeared to be a largely theoretical possibility (Meadows 1972), but now would appear to be real (Meadows et al. 2004). Wealth creation would now seem to be limited by the dwindling supply of natural and human resources. This has a number of possible consequences for children depending on policy maker's choices and particular children's position with respect to 'standard' childhood, among other factors. Here are three illustrative possibilities.

A. The suggestion of tougher times ahead may increase governments' and individuals' determination that their children will be as well prepared for a future of intense resource competition as possible. In this scenario, existing means of boosting children's future ability to contribute to national and personal economies, such as schooling, may be used more intensively. New means may also be sought. Arguably, recent advances in genomics and neuroscience are beginning to provide those new means and a growing global middle class may provide a market for them. In this world a few children would be 'enhanced' with or without their knowledge and consent, while many others would struggle for food and water.

B. Alternatively, the relatively narrow definition of progress and well-being in terms of economic growth that has been so influential throughout the twentieth century may well be broadened to involve a wider range of human capabilities than wealth creation alone. In this scenario the primary goal of education as the development of the cognitive powers of the young may increasingly be joined by other priorities and age-defined targets such as the ability to manage one's own health, happiness or environmental sustainability across the life-course.

C. Thirdly, if current forecasts are correct, we are likely to see increasing displacement of populations either as a direct result of climate change or as a result of consequent conflict for resources. This will pose questions of what can and should be done with the many children involved in such population flows. Starkly stated, which of them will be seen as an investment opportunity and which will be allowed to die?

Each of these scenarios exceeds the existing terms of social studies of childhood. They hinge on such topics as genomics, neuroscience, the resource implications of children's lives and sovereignty over life and death. Given all this, the by now well-rehearsed view of childhood as a socially constructed 'form of life' needs to be supplemented with an understanding of children as 'life-forms' (Rose 2007). As I have argued throughout this chapter, there is no suggestion here of return-ing to a view of childhood as a universal, purely natural condition. Indeed, a critical issue will be to demonstrate that seeing children as 'life-forms' will allow us to see variation alongside commonality and adaptability alongside understandings of natural regularities in chil-dren's growth. Rather, the focus will be on presenting and illustrating the evidence and concepts needed to adapt childhood studies in the light of current trends.

2
Childhood and Bio-politics: Life, Voice, Resource

In the previous chapter I traced connections between the present agendas of childhood studies and a late twentieth-century wave of economic, cultural and legislative globalisation. I argued that the contemporary focus on the cultural diversity of childhoods and on the question of children's self-representation emerged in response to those broader social and economic developments. I then highlighted some more recent developments in the life sciences and in understandings of climate change. These developments suggest to me that a strategic reframing of the basic questions of childhood studies would allow the field to grow and adapt in response to these changes in turn.

I then drew attention to some common commitments of the field; the use of a binary frame (nature/nurture, nature/culture); and, an understanding of life processes that tends to reduce their enormous diversity to a unitary 'nature'. I suggested that these commitments, if maintained in coming years, would limit the relevance of childhood studies in a world that, in significant part, no longer shares them. I introduced some researchers who have tried to register these changes and respond strategically to them in other fields and suggested that childhood studies can learn from them. At the same time, however, I cautioned against abandoning the stock of sensitivities, skills and practices that the last hundred years or so of childhood research has built up. In this book I certainly am working towards change in the range of frames that inform and organise childhood research, but I also have pronounced conservative impulses. Simply stated, the risk of adaptation is of losing rather more than you gain.

Just as 'new paradigm' (James and Prout 1997) and 'social constructionist' (Burr 1995) approaches tended to lose touch with children's
creaturely existence as they turned away from life processes, so shift-
ing away from a binary frame that has helped structure research for
many years could also involve loss.

In my view, then, a binary frame is not to be understood as if it
were a factual or theoretical error that could or should simply be dis-
avowed. Rather, binaries of nature/nurture, nature/culture and being/
becoming are best understood as strategic resources that have seen a
lot of use in formulating theoretical issues, empirical questions and
ethico-political positions within specific contexts. Each set of paired
opposites has been used as if it were a compass indicating directions
of travel and allowing for the navigation of research fields, just as real
magnetic compasses help divide North from South. As new contexts
emerge, the challenge, as I see it, is to produce new sets of resources
for navigating and organising research and to do so in a way that
preserves connection with the history of childhood research.

This chapter takes on this task of assembling new navigational
and organisational resources for childhood research that can both
complement the history of childhood research and orient present-
day research to children as 'life-forms' (Rose 2007). In the previous
chapter I spent some time sketching Foucault's concept of 'bio-
politics'. Building these new resources will begin with further detailed
description of 'bio-politics' as a field of social and governmental
action, considering it in relation to key turns in the story of childhood research. I will use Foucault (2007, 2008) and Agamben (1998),
the two key authors on bio-politics, to put the binary frame around
childhood into the historical context of its development. This should
clarify how it came to be useful. As I go along it should become
clear just how the binary frame helped childhood research play a
part in bio-political governance. The fact that scope already exists
for productive new ways of navigating and organising contemporary
childhood research should also gradually become apparent. For now,
I will simply note that I will be trying to supplement the binary
frame with an awareness of the key 'multiplicities' of knowledge
and of practice that have emerged through a history of strategic
and tactical engagements between states and populations to structure
childhoods over the years. I call them 'Life', 'Resource' and 'Voice'.
Where the binary frame helped navigate the uncharted space of

'human nature', these multiplicities allow for connections and comparisons in a complex and emergent, but far less mysterious, space – the material world in which designs, desires and life processes meet and mix. So first, here is a more detailed examination of Foucault on the bio-political.

States, populations and security

In the previous chapter, I spent a little time describing Foucault's use of the term 'bio-politics'. I indicated that, in his view, 'bio-power' emerged in eighteenth- century Europe to supplement existing 'juridical' and 'disciplinary' forms of power that were already in play in relations between government and governed. Juridical power treats humans as a set of legal subjects who are capable of voluntary actions for which they can be held responsible. Disciplinary power addresses humans as bodies that can be trained to perform certain actions in a required way. Bio-power, in contrast, addresses humans as a species, as a population of creatures that are capable of autonomous action and have certain behavioural tendencies. This population is always closely bound up with and reliant on the characteristics of its immediate environment or 'milieu'. Bio-political strategies are distinguished by their use of the characteristics of a given milieu and of humans' own behavioural tendencies as tools to act on the population so as to increase the security of that population against such threats as disease and starvation.

Across two closely argued books, Foucault develops the view that the many historical successes of bio-political strategy have come to create a contemporary relationship in which states tend to view their populations, especially their young, as a resource, as 'human capital' (Foucault 2008) to be developed and used in the interests of general security. Bio-political power involves states in a commitment to make the most of their population's capabilities and behavioural proclivities through strategies that are designed, as far as possible, never directly to force a populations' compliance with policy, but rather to work with the 'grain' of human behaviour, to seek paths of least resistance; in short, to govern in line with the dictates of 'human nature'. For Foucault (2008), it is on this basis that contemporary 'neo-liberal' forms of government can be understood as liberating individuals even as they govern populations. The liberation of natural energies, desires and proclivities is presented as the very means of government. It is important to note that the development of this approach to government was not stimulated by the discovery of indisputable facts of 'human nature'. It is not an example of evidence-based policy. Rather it was governments' desire to use people's behavioural proclivities within their milieux as a resource in the pursuit of security that set scientific research in psychology, sociology, economics off in search of 'human nature'. It was in the course of this effort that the poles that in large part still define childhood studies (nature/culture, nature/nurture) gained their function and credibility as navigational devices.

I'll take some time now to unravel the terms 'milieu', 'population' and 'security' using two of Foucault's historical examples. But I'd like briefly to note at this early stage that in Foucault's analysis, bio-politics contains tensions that might make those of us who are ruled through it feel uncomfortable. First, we might be uncomfortable thinking of ourselves as a 'resource' or as 'human capital'. The terms contain a moral ambiguity. On the one hand, we might think of a 'resource' as something that can be used without being consulted, like a sack of coal, and we might want to avoid being characterised in that way. On the other hand, if we understand and value the purposes we are resources for, we might proudly volunteer to be 'used'. Second, under bio-power we are used as resources for something approximating our own good, or at least for the general security of the population. The experience of being told that something dull or disagreeable is 'for your own good' may be familiar to you from your childhood. It is a more significant experience than it might at first appear. Children have long been and continue to be a principle focus of bio-political strategy. Thus, children are often positioned as not knowing what is good for them, while others, usually adults, are supposed to. The key contemporary challenges facing bio-politics of human potential, well-being and ecologically sustainable living that I will describe in later chapters involve dealing with precisely these tensions and sentiments.

Strategies against scarcity

Foucault describes 'bio-politics' as a new set of techniques of government that were developed in the seventeenth and eighteenth centuries to address the novel challenges of the times. By the seventeenth century, much European trade and administration had become concentrated in towns. Even as the milieu of the town brought certain efficiencies of communication with it, relatively high population densities also brought problems. Infectious disease such as smallpox and meningitis (Crawford 2007) could spread easily through close human contact, as could political revolt. A principal spur to revolt was the threat of scarcity in staple foods like wheat and other grains. Wherever a population feared the scourge of mass starvation, governments also feared for their position. Thus, a great deal of attention was given to securing the populations of towns against grain scarcity. Foucault draws a clear transition in the development of such strategies over the seventeenth and eighteenth centuries from the early use of juridical strategies that sought security in the prohibition of certain behaviours to the later use of distinctively bio-political strategies that relied on 'human nature' – the perceived characteristics of humans as a species – as a tool and a resource of government.

Juridical strategies against scarcity tended to be based on moral judgement about human behaviour. They were organised around the identification and prohibition of acquisitive or greedy behaviour on the part of those who produced or owned grain. The main strategy was to prohibit any behaviours that were likely to raise the price of grain. Thus, in order to ward off scarcity, strict limits were set to the quantities of grain that could be stored, so that hoarding behaviour, intended to drive prices up, could be prevented. Likewise strict controls on the export of grain across national boundaries were put in place to prevent grain producers from maximising their profits by, effectively, denying their immediate neighbours' needs.

These measures certainly kept grain prices low, but were less effective at preventing scarcity than we might expect. Low prices limited peasant-farmers' motivation and practical ability to invest year-on-year in grain production. If returns are limited in advance, why go to the trouble of draining and ploughing a new field? If returns over the last few years have been modest, how could you afford the labour or the draft animals involved? According to Foucault, this system could actually increase populations' vulnerability to scarcity. A relatively small and under-capitalised agricultural system could not produce much in the way of surplus grain to act as a buffer against a poor growing season; anti-hoarding measures ensured that such surpluses could not accumulate; and, international export controls meant that a starving nation would find it very hard to source grain from neighbouring countries. Juridical strategies identified certain human behaviours as wrong or sinful. Think of the figure of the grain hoarder refusing to sell till people are desperate. Thus, they treated scarcity as if it were as rooted in bad behaviour and sought security in attempts to prevent that bad behaviour. These strategies drew a close connection between individuals' moral character and the phenomenon of scarcity, almost as if sin were the cause of starvation.

In the eighteenth century a fresh approach emerged that had little time for moral judgements about individual behaviour but instead sought deliberately to deploy behavioural tendencies that would once have been denounced as sinful in a positive manner as tools in the prevention of scarcity. Greater permission both to store grain and to trade it internationally was given. This certainly enabled grain producers and merchants to accumulate greater wealth because it gave them a stronger hand in price negotiations. But by increasing the geographical spread of trade and lengthening the period of time between harvest and sale it also set these traders in closer competition with one another. Here then, the self-interest of grain producers in competition with one another could be relied upon to do the bulk of the work of stabilising grain prices and securing grain supply. Rather than judging and condemning individual behaviour, government could focus on ensuring that the many flows of goods and information that grain markets relied upon were smooth and on increasing the population's confidence that grain would always be available. The upside of this new approach, both for towns' populations as a whole and for government, was clear. The threat of mass starvation and political revolt receded. But there was a downside too. Juridical strategies had kept grain prices low. Thus, whenever grain was available the vast majority could afford to pay for it. Bio-political strategies, however, allowed prices to fluctuate. This meant that some individuals went hungry or even starved amidst the general plenty of the population. Thus, bio-political strategies created securities for populations and for states by re-distributing poverty and suffering from occasional mass starvation to semi-permanent scarcity for a few.

Government through human nature

In the transition that Foucault (2007) charts, the key issue for government changed. It was no longer focussed solely on how to prevent individual sin and the disastrous consequences of sin. Instead a parallel concern had developed to understand the behavioural tendencies of a population so that they could become tools and resources in the production of security for that population. In the sphere of grain markets at least, government became about working 'with the grain' of human nature. It is important to note that this fresh sensitivity to 'human nature' was not the result of a great scientific advance of the eighteenth century in which fundamentally new facts about humans were discovered. Rather, in the context of grain supply, this 'human nature' was simply acquisitiveness and self-interest, hollowed out of moral significance and re-cast as a potentially useful set of behaviours.

A number of themes of great significance for contemporary childhood studies are woven into this transition. First, it is here that government on the European model began its journey towards thinking about the population as a resource which, having its own preferences, proclivities and capabilities, can be worked on to produce security and stability. We can, for example, chart a continuity between Foucault's eighteenth-century grain markets and the workings of the Organisation for Economic Cooperation and Development (OECD) which makes international comparisons of the relative economic development of countries by assessing educational systems alongside the flexibility of the labour market and a host of other factors. In many countries, children and young people, figuring as human embodiments of the future, are a particularly intense site of application for the logic of human capital and security. Second, is the question of 'freedom'. A purely juridical government faced with a major challenge to security would simply try to impose its will on individuals. Bio-political government however needs to preserve the autonomous preferences and proclivities of populations if it is to carry on using them as resource. Working with rather than against such preferences and proclivities relieves government of the requirement to micro-manage individuals' lives and actions. It also preserves the legitimacy of government since it tends to strengthen the perception that the population is governed by its own consent. These two issues are drawn tightly together wherever 'human futures' are under discussion and policies created to secure states' and populations' futures through them.

None of the points above imply that Foucault's bio-political strategies are all about granting liberty to some category of 'natural' human impulses. Even though he describes the dominant contemporary style of bio-politics as 'neo-liberal', it only appears 'liberal' or 'free' in comparison to those juridical governmental strategies that sought to prevent selected behaviours on moral as well as practical grounds. Ideas of freedom in flows of goods and people actually represent a fundamental tension of bio-politics. On the one hand, behavioural tendencies of the human species need to be recognised without making immediate moral judgements about them. On the other hand, for bio-political rule to be successful in producing security, populations need to be understood in detail so that appropriate interventions can be designed that make the most of a population's range of natural capabilities and proclivities. Bio-politics requires that capacities and limitations of 'human nature' be discovered and worked on so that they can act as a reliable basis of policy. This can be illustrated with a brief consideration of the second major security threat of the eighteenth-century European town – epidemic disease.

Disease and population

Smallpox is a highly infectious viral disease that has given rise to epidemics for as long as humans have lived in towns and cities. An epidemic could kill as many as a third of an affected population. Fortunately for us, the virus was eradicated in the late twentieth century, save for a few scientific samples, but in the eighteenth century it was a very real threat. For example, in eighteenth -century London, epidemics broke out every five or six years, infecting and killing individuals across social strata. Traditional European treatments for individuals who had contracted smallpox, such as using heat to 'sweat' the disease out, leaches to 'bleed' it out or dressing the patient in red clothes were, perhaps unsurprisingly, ineffective (Crawford 2007). However, for hundreds of years in India and China, a practice known today as 'variolation' had been used to minimise individuals' susceptibility to infection. Dried smallpox scabs were collected from sufferers, powdered and puffed up the noses of children. They then contracted a relatively mild form of smallpox from which most made a full recovery, followed by immunity to the disease. A few children would die from this procedure but far fewer than would die if left untreated in epidemic conditions. Variolation was brought to Europe and the United States from Turkey in the 1720s by Lady Mary Wortley Montague. She had lost a brother to the disease and had been scarred by it herself. She had her family doctor perform the technique on her children and their survival and subsequent immunity encouraged other aristocrats to do the same. Variolation against smallpox then became common practice throughout the United Kingdom.

Many medics of the time were actually opposed to variolation. There was no basis for the practice in the medical theories of the time; it seemed unwise deliberately to infect children risking not only their individual deaths but potentially sparking an epidemic. Further, the practice had been introduced by a woman. Some argued that smallpox was best understood as a divine retribution for sin, thus that attempts to prevent it opposed the will of God. Foucault (2007) accounts for its popularisation in the face of these obstacles in terms of the transition between juridical and bio-political governmental strategies. Not only was variolation simply effective, but it also shared key features with emergent bio-political strategies against scarcity. As I described above, new means of preventing scarcity depended on breaking the link between perceptions of human nature and moral judgement and, instead, finding ways to make positive use of people's capabilities and proclivities as tools in the pursuit of security. This involved a change in the distribution of suffering that secured the population as a whole at the expense of some individuals. Occasional famine striking whole populations was replaced with the near-permanent food poverty of those individuals who could not keep up with fluctuating grain prices. Likewise variolation involved breaking the link between disease and divinity. Instead of seeing the human body as the register of God's will, variolation, in effect, made use of the capacity of the human body to adapt its immune system in the light of exposure to infection. As more individuals were treated, so the threat of smallpox epidemics affecting entire populations receded even as a small percentage of children treated would inevitably die. Once again, this relationship between security and population is not about letting the characteristics of humans as a species simply run free – whatever that would mean. Rather, it is about relying on 'human nature', seeking a point of support in the capabilities and proclivities of humans as a species, such as the desire for wealth or the responsiveness to infectious agents, and, '... instead of trying to prevent it, making other elements of reality function in relation to it ...' so that threats to security are cancelled out. (Foucault 2007: 59).

Bio-politics, childhood and the binary frame

Foucault's concept of bio-politics is an attempt to summarise the principles of a range of governmental strategies that underlie the forms of governance that link state and population in many regions today. It is one part of his broader attempt to understand how, over the course of a few hundred years, a consensus has emerged among many, if not all, governments that their fundamental task is to secure the well-being of their populations and that their principle resource in doing so is the set of capabilities and proclivities that make up the 'human nature' of their citizens. When considering the concept 'human nature' or Foucault's preferred 'human species', it is important for us to remember that this was not formed on the basis of a positive scientific discovery of some set of objectively determined facts about us as a species. After all it wasn't until the nineteenth century that ideas about human evolution began to be formalised in scientific practices (Darwin 2003). Rather, the path to thinking of humans as a species with certain behavioural characteristics began with a set of moral characteristics – sin, virtue, good and bad behaviour – emptied them of moral significance and turned them into tools of government in the service of the practical objectives of preventing scarcity and epidemic and other threats to security. Foucault suggests, then, that the historical development of strategies of governance both preceded and informed the development of the human sciences including those that focus on children. This is helpful in understanding how childhood research has been able to make such successful use of a binary frame that depends on the referent 'nature' to define its major questions (nature/nurture, nature/society, nature/culture) and to find a niche for itself in the marketplace of ideas.

In bio-politics, Foucault presents us with a style of governance that would seem to confound a binary frame. For example, we might normally think of issues like desire for wealth and perceptions of self-interest and the functioning of the human immune system as belonging to very different areas of thought and knowledge. The first seems rather 'social', the second 'biological'. But it is clear that if Foucault is right about bio-political rule, then, from the bio-political perspective, desire for wealth and immunity to infection are just different versions of human natural resource, of the autonomous capabilities and proclivities of humans as a species. From the perspective of bio-political governance, the social/biological distinction is not of fundamental significance. The two cases of desire and immunity certainly differ in terms of what sort of interventions are likely to affect them. For example, it is probably no use trying verbally to persuade a virus that it is not in its interests to infect an individual. But their openness to interventions will sometimes overlap. For example, international trade regulations may well influence the global distribution of pathogens alongside patterns of investment (Crawford 2007). If this bio-politics is indifferent to the social/biological distinction, however, and if as I claim, the study of childhood has been enabled and shaped by bio-political concerns with human nature, why has a binary frame been so useful over the years for childhood researchers?

As I have argued, Foucault's account suggests that the 'human nature', so fundamental to bio-political strategy, was formed from a list of human capabilities and proclivities originally identified as sites to apply moral judgement. The key innovation of bio-political strategy was to take those same sites, but to withhold moral judgement about them so as to open them up to use as governmental tools. Key European religious and philosophical traditions of moral thought had long had a special place for children. They were often positioned as being particularly close to the fundamental truth about humans. Jenks (1996) compares Puritan and Romantic conceptions of childhood. The former, for example, depicted children as bearers of 'original sin', while the latter privileged them as peculiarly virtuous. As bio-politics re-coded human behaviour as a matter for governmental analysis rather than divine of individual judgement, children retained their special, if unenviable, position as closer in thought than other humans to the truth about the species. This is what established the space for childhood research and made a binary frame especially useful as that tradition developed in the twentieth century.

If government was to rely on humans' autonomous capabilities and proclivities and to do so in independence from moral judgement, it needed to know about these. The younger the human, the less time moral judgements will have had to distort their capabilities and proclivities. If the bio-political truth of the human is an 'autonomy', a 'spontaneity' conceived of as 'natural', then children appeared to be an ideal case for study. Further, if populations are a resource, then determining the full range of natural resources present in the child and discovering how best to foster and utilise them are issues of great importance. The most effective pathways to changing the expressions of this spontaneity also need to be discovered if human capabilities and proclivities are to be set against each other in such a way as to secure the future for a population. Do we expect the future to be predictable? If so, how are we to encourage behavioural conformity in the young? Do we expect the future to be unpredictable? If so, how are we to engender resilience or an entrepreneurial outlook? Practical political questions like these make curiosity about the degree and means of manipulability of human resource vital. The tensions between nature and nurture and between nature and society that informed so much twentieth-century developmental psychology and sociological study of socialisation have, thus far, been the most prominent expressions of this historically and politically specific form of curiosity.

Complexities of childhood bio-politics

The conditions of emergence of bio-political strategies and their relationship to childhood research should by now be clear. It is important to emphasise however that, for Foucault, the bio-political is just a set of strategies that emerged under specific conditions – the milieu of the town for example – to address specific problems – scarcity and epidemic. As I have suggested, the basic bio-political relationships between states, populations and future security established years ago have seen a lot of success and have mutated over the years into a very clear identification of populations, especially children, as 'human resources' (Foucault 2008). But this relatively simple account should not be taken to indicate that today's bio-political rule is complete and monolithic or even that it is always successful, even in its own terms. Foucault makes it very clear that bio-political strategies against scarcity usurped juridical strategies in part because the latter were not successful in preventing scarcity. Similarly, many of the topics I will discuss later in the book, such as climate change and sustainability, have arisen in spite of and in some cases because of specific bio-political strategies. Examples here would include the production of economic growth through the promotion of desire in societies of mass consumption.

The scene of government as it relates to children remains a complex and unpredictable one. Even if the main script for government today is bio-political, juridical and disciplinary strategies of judgement and training are still deployed at local levels, often to assist bio-political ends. Training children to see themselves as responsible for their own actions and insisting that adults see themselves that way are still major political strategies. But even the prohibition of conduct on moral grounds is inflected with bio-political strategy in complex ways. In one contemporary style of child discipline, for example, it is considered essential when correcting a child that it is their behaviour that is 'bad' rather than themselves (Pantley 2007). Similarly the UK criminal justice system attempts to reconcile distinctively juridical concerns with moral accountability and bio-political concerns for the best development of young offenders. This tension informs both specific controversies over sentencing of 'guilty' children and the general question of the age of criminal responsibility (Cipriani 2009).

Bio-political rule generates fresh problems of government even as it resolves others. Further complexities are added by the key tensions inherent in bio-political rule I listed earlier. Being positioned as a resource can be offensive and threatening, even or especially when this implicitly done in one's own best interests or in the name of one's liberation. Thus negotiation and bargaining between governments and populations about children as a security resource is a constant in contemporary bio-political rule. It becomes most obvious when it breaks down. For example, it is thought that over the last decade in the United Kingdom levels of vaccination against common childhood diseases measles, mumps and rubella have dropped significantly (Cockman et al. 2011). In 1998 a publication in the prestigious medical journal *The Lancet* reported evidence that the, then standard, MMR vaccine that dealt with all three diseases in one go was associated with bowel disease and autism (Wakefield et al. 1998). The claim was made that some children's bodies were overwhelmed by receiving all three vaccines in at once precipitating neurological developmental disorder. This view has now been conclusively discredited, but some parents' withdrawal of their children from MMR vaccination on the basis of that report tells an interesting story about the complexities of bio-political rule.

Recall that variolation against smallpox killed a few children but preserved far more. Today's vaccination science usually ensures that the suffering of the few for the benefit of the many is minimal. The proven risks of MMR to children include fever, rash and a degree of joint pain with a very low incidence of serious allergic reaction. The pronounced dip in vaccination rates following the *Lancet* publication indicates parents' high sensitivity to the balance between their child as an individual susceptible to side-effects of vaccination and as a member of a population that tends to host communicable diseases like measles, mumps and rubella. It is clear that the notion that authorities are taking risks on behalf of the population that individuals would not themselves choose is a strong theme of the bio-political 'subconscious' around children's health, in the United Kingdom at least.

It is tempting to criticise contemporary forms of bio-political rule on the grounds that they contain 'contradictions' between apparent government diktat and individual autonomy. Some critical commentators on contemporary childhood do exactly that (Furedi 2002), while others prefer to decry the mobilisation of children's choices and freedoms (Seligman 2009). But this would be to assume that the bio-political strategies that surround and help compose childhoods aim at ideological coherence as if they were political parties trying to define a consistent message to present to voters. In my view, Foucault's most important contribution is to recognise that the bio-politics of relations between states and populations is, at root, a style of framing and responding to perceived problems – quite what the problems are, where they come from and what events provoke them vary enormously with both context and perspective. Sometimes those who are developing strategies give emphasis to populations as a resource of capability and proclivity for government to shape – as in a vaccination campaign. At other times the stress is on enabling the autonomous expression of proclivities and preferences – as in the removal of barriers to international trade and competition that spurred economic globalisation. Either approach in any combination is conceivable as long as it provides 'security', however that is defined. Just as 'human nature' remains controversial within bio-politics so do definitions of 'security' and of the best means to achieve it. It is often the case that one version of security wins out over others. In Chapter 4, for example, I describe the current dominance of the concept of 'mental capital' with its emphasis on economic competition and its implications for cementing some rather limited definitions of effectiveness in children's learning.

Foucault, Agamben and alternatives to the binary frame

Foucault's is not the only voice on bio-political power. The political philosopher Agamben (1998) has been strongly influenced by Foucault but takes a rather different approach. He thinks that bio-politics have a much longer and more complex history than Foucault (2007) describes. Agamben traces a development in the relation between nature and politics all the way from Ancient Greece of the third century BC to the present day. Foucault's central question is how the view of populations as 'human capital' with all its implications for intervention in ways of life came to be so widespread in the contemporary world. Agamben's concern is with a certain loss of measure and discrimination in political thought and action that is certainly related to 'human capital' but extends beyond Foucault's core concerns critically to address the idea of 'human rights' alongside today's international political order of well-being and resource scarcity, migration and humanitarian military action. As he describes this concern, he adds many insights into the strengths and weaknesses of a binary frame.

As I've suggested so far, childhood is often approached with a binary frame. One of the things I want to do here is show that it is possible and productive to approach it otherwise. This involves three stages. First, I need to show that such distinctions are not inevitable even if they have a long history. Second, I need to show that even though the binary frame is not inevitable, it can still be effective in structuring research and positioning it in the marketplace of ideas.

I hope that my coverage of Foucault has established these points. The third stage is rather more difficult. It is relatively easy to establish that a given way of thinking is not inevitable and to envision the possibility of alternatives. Then it's just a short step to deciding that, if our categories are not as reliable as we once thought they were, what we really need to do is abandon such categories entirely, perhaps collapsing them into each other on the grounds that they do not really reflect the world we live in. But the next step I want to take is to show that if we simply do away with distinctions we can end up worse off than we were before. To my mind, the distinctions involved in the binary frame are tools. This means that if we want to discard them but want to carry on analysing contemporary childhood, we probably need to make some new tools. One way of reading Agamben's views on modern bio-politics is as a cautionary tale of what can happen if distinctions are discarded and not replaced with others.

Law, nature and sovereignty

The ancient Greek philosophical and political traditions that have influenced many contemporary societies contain a distinction between two kinds of 'life'. The Greek words 'zoë' refers to 'life' in the sense that all living things, from grasses to whales to people, have it. The word 'bios' refers to 'life' lived in the light of public standards and expectations. It names the political life that is proper, in the ancient view, to people who live in cities. The distinction between these two kinds of life was an important part of a set of philosophical questions about whether and how people differed from other animals, but it was also of fundamental political importance. It provided a frame that was useful in addressing a set of political problems.

Imagine yourself as the ruler of a city of about 250,000 people. You are responsible for making the laws about how people should conduct themselves, for preventing conflict and dealing with it when it arises, for dividing up resources that are always limited, and for distributing punishments varying from fines to executions. The life, death and well-being of the population depends, in no small part, on how you govern. To help you enforce order, you have a militia at your disposal, but they are not a large enough force to crush a widespread rebellion against you. So how do you ensure that your rule is seen as legitimate? When the justifications for your sovereignty, for your rights over life and death in the city are unclear, there is at least one thing you can do with complete certainty. You can establish a 'frame' around your powers. You can declare the limits to and identify the exceptions to your powers. For Agamben, bio-politics begins with this matter of exception dividing zoë from bios. This frame, this system of exceptions, still informs our world. There are some things that law can and should cover. These are matters of value about which it is proper to make judgements of right and wrong. But there are other matters to which, by common consent, decrees and legal codes cannot or should not be applied, the matters of fact that make up the natural world.

On Agamben's view, it is this self-limitation of sovereignty that provides sovereignty with legitimate foundation. It promises citizens that, in general, there will be areas of the life of the city that the sovereign will see as 'zoë' and will not try to intervene in. Clearly, a wise sovereign will not try to use legislation to make a goat to produce more milk. But sovereign self-limitation extended further than this. Domestic life and reproductive labour of childbirth and childrearing were also originally exempted as zoë from the sovereign's power and left to the control of the father of the family. Sovereign self-limitation also promises citizens that, in general, the sovereign will not use his power to take away their life in an arbitrary fashion. Nevertheless, the very structure of these promises means that the sovereign always retains the power to decide where and when exceptions can be made to these general promises, where and when the rule of law can be suspended and supplanted by unaccountable, lawless responses.

The exception becomes the norm

As I have suggested, the zoë/bios, nature/law, life process/life style or, indeed, fact/value frame result from the self-definition and self-authorisation of sovereign power by self-limitation. Drawing the distinction between nature and law, between what is included in sovereign power and what is excluded from it is, for Agamben, the founding move of bio-politics. It is a strange and paradoxical distinction however. Though it divides those parts of the world that belong to 'nature' from those that demand public legal regulation, the distinction itself belongs to neither side. Just as Foucault indicates that the 'nature/nurture' frame precedes the human sciences, Agamben points out that there are no natural laws that underlie the lifestyle/life process distinction or call it into existence. Likewise, since it is the very foundation of legal legitimacy, no law could possibly be said to authorise it. In one sense, then, it is an entirely arbitrary distinction. In another sense it is fully necessitated by the vulnerability of any figure foolish enough to claim unlimited sovereignty. Think, for example, of the absurdities captured in the fable of King Canute who was supposed to have ordered the tide to turn back on itself.

Just like the measures for managing scarcity and disease that Foucault described, the zoë/bios distinction is a political strategy for managing the vulnerability of power. It demands that the sovereign, who has power over life and death, accepts and states general limits to that power. But, having no grounds other than itself, it also gives the sovereign the power to decide on exceptions to the general case, to include what was once considered 'zoë' into the expectations of law and public standards or to decide in a given situation that the rule of law can be suspended to reveal erstwhile citizens as mere 'zoë'. Agamben's central example of this is the ancient figure of the 'outlaw' over whom exceptions to normal respect for life can be suspended such that he may be killed without ceremony and without subsequent punishment for his killer. Lodged in private domestic space, the lives and well-being of the many women, children and slaves of Greek antiquity who were dependent on the family patriarch were in parallel, if not identical, exceptional circumstances with respect to the sovereign's public law.

Ancient Greece is not really so far away from us today. Unchecked, extra-legal patriarchal rule is a fact of life in many parts of the world. It is often referred to as 'tradition'. Eradicating this exception is a major motivation behind bids to 'modernise' and develop nations by, for example, providing girls with schooling in Afghanistan (Mortensen 2010) and increasing women's control over their own fertility worldwide, all in the interests of health and well-being. But for Agamben, the world of politics has seen significant variations on the theme of life, rule and exception over the centuries. In his view, the contemporary bio-political scene is best understood as a situation in which the exception has become the rule, where governments all too often rule by making exceptions to the way they would normally categorise people as citizens and as resources.

Recall how in Foucault's account today's bio-political relationships between states and populations are now organised around the idea of people as instances of a species with certain capabilities and pro- clivities. For Agamben something big did happen in the eighteenth century. But it was not the emergence of bio-politics against a back- drop of juridical and disciplinary power as Foucault has it. Rather it was the increasingly deliberate and programmatic involvement of states in the very 'human nature' or 'zoë' that, in the terms of classi- cal political philosophy, would normally be excepted from sovereign rule. As more and more political programmes focussed on humans as resource, the kind of involvements that had once been exceptional gradually became the norm. It is now normal, for example, for gov- ernments to immunise their populations against infectious disease or to have strong role in shaping the capabilities of future adults by providing access to schooling, taking firm hold of children as human fragments of the future.

Abandoning a binary frame?

Agamben portrays the current bio-political scene as one in which a state of exception has been generalised. In the past, he argues, states defined themselves by declaring limits to their rule. This often meant distinguishing between the 'life' (zoë) that humans share with all other creatures and the 'life' (bios) that is peculiar to citizens that is lived in the light of publicly recognised standards. But since 'human nature' rather than 'sin' became the key focus of political rule, the sit- uation is that, in principle, everything and nothing can be excepted. Under these conditions the distinction between zoë and bios tends, in practice, to evaporate, even if we often still talk about biolog- ical and political life as if they could be clearly demarcated. It is at the point in his account that Agamben gives us cause to ques- tion abandoning our binary frame unless we also work hard to find replacements. If we carry on talking and researching as if biological and political life were cleanly separated, we will miss a lot of the important action in contemporary governance. Alternatively, if we participate in this abandonment of the binary frame, of the zoë/bios distinction without finding some replacement, we will lose the abil- ity to hold the powerful, such as our national governments and the international organisations they support, to account.

For Agamben this means that while the Ancient Greek paradigm of political rule was the 'city', in the modern world a more suitable model is the 'camp', in particular the concentration camps of the Nazi regime. Consider that a primary purpose of the Nazi concentration camp was the elimination of people on the grounds of their Jewish ethnicity. In the Nazi state 'Jew' and 'Aryan' were understood as biological categories – that is to say as the product of variation in human nature. The application of racist ideas of biological difference in human nature allowed for matters of human embodiment to be entirely given over to definition in political terms. Biological, cultural and moral identities were all mixed together in Nazi theory and propaganda. Rather than seeking legitimacy in the distinctions of the binary frame, the Nazi state deployed an indiscriminate blend of statements about politics and about nature. That lack of distinction meant that even though Jews were detained in concentration camps under Nazi law, their lives could be exempted from law, exposed as bare life and killed without ceremony or sanction as you might exterminate an insect. A stark indicator of the generalisation of the exception in this context is that the Nazi state required of itself that prior to their murder, Jews be formally 'de-nationalised', in other words, defined as outlaws (Agamben 1998: 132).

For Agamben then, the Holocaust is a prime example of what can happen when the state of exception becomes the norm. He also invites us to consider certain other contemporary arrangements as based on exactly the same structure of indistinction. He offers us the idea of human rights to life, humanitarian responses to majority world famine, poverty and disease and consumer capitalism. He seems quite aware that accepting these structural parallels will involve a considerable stretch of the imagination for some readers. As he puts it, '... the camp will appear as the hidden paradigm of the political space of modernity, whose metamorphoses and disguises we will have to learn to recognise' (Ibid: 123).

Responding to Agamben

My own response to Agamben's argument is quite ambivalent. On the one hand, drawing a parallel between the Holocaust and consumer capitalism, for example, seems, at the very least, heavy handed – a case of using a sledgehammer to crack a nut. On the other hand, contemporary consumer capitalism, for example, does function by stimulating and maintaining desire for its products and does so by appeal to what might be understood as people's more creaturely aspects. Sexualised imagery and the imagined comforts of food and home are all used to sway consumer choice. At the same time, certainly in Anglophone polities like the United Kingdom and the United States, consumer choice is being mobilised as a key mechanism for political decision-making. This looks like the indiscrimination Agamben warns of. The non-governmental organisations (NGOs) that organise humanitarian responses to refugee crises and famine often struggle with the difference between short-term life-saving solutions and longer term improvements in political and economic relations. This is because they often work within sites of exception where states either cannot set legal and welfare standards or have withdrawn from doing so. They sometimes have to work hard to maintain the aura of political neutrality that can be essential to their life-saving work. Nevertheless, the argument can be made that humanitarian responses to the suffering brought about by war and famine, where they are organised around saving lives tend to simplify and depoliticise the contexts in which they take place (Barnett and Weiss 2008).

Life, Voice and Resource: Multiplicities of childhood

In this chapter, I'm certainly trying to step to one side of the binary frame but also to preserve the means critically and creatively to engage with the many ways life processes and lifestyles are drawn together around children and childhood. So far, I have examined the two key authors associated with the term bio-politics. Foucault's insights helped to place two-category thinking associated with the binary frame in a historical context. This is a reminder that no matter how much we take nature/society and nature/nurture divisions for granted, and even though they have shaped scientific approaches to childhood, they are not part of the structure of the world. In other words, none of us are under any final obligation to use them to structure our understandings, expectations and practices. There is no reason at all why we should not devise other ways of organising our research and thought. Yet the binary frame has clearly been useful for some purposes. Where the task was to best deploy populations over time in the interests of keeping most of them sufficiently healthy and wealthy, figuring out what capacities they have and how those capacities might best be shaped and harnessed made sense as a set of practices. So how might we capture the role and functions that the binary frame has had? The best metaphor I have been able to come up with is a navigational one (Lee and Motzkau 2011). The twin poles of two-category thought were, in practice, used like the poles of a magnetic compass to orient childhood researchers. It gave a sense of what the important questions in the field were – often whether and to what extent children can be shaped toward a desired future or towards a norm. Further, wherever those futures or norms became a matter of political controversy, say in the late twentieth century over the socialisation and development of gender roles, the compass needle helped researchers define their own positions, identities and motivations for research. For all its drawbacks then, the binary frame often makes sense, given the right context.

Having learned from Foucault, I then turned to Agamben. He tries to establish the idea that when it comes to thinking about and researching human existence, distinctions matter. He is willing to make comparisons between the appalling contortions of Nazi self-legitimation in concentration camps, where humans were defined simultaneously as entirely political beings and as entirely animal beings, with contemporary consumer capitalism in which an impulse buy can readily be understood as a form of political participation. He does so to show how forms of rule that systematically efface distinctions between zoë and bios effectively insulate themselves against critique. One way to understand his willingness to make controversial comparisons is his desire to retain the ability to make analytic distinctions in the face of such figures of self-sealing legitimacy as consumer capitalism, humanitarian aid and, even, human rights. When questions about the wisdom of the use of massive personal wealth by individuals to shape a regions' health policy can be silenced by reference to the lives being saved (see Chapter 5) and when attempts are made to base the legitimacy of all national and international politics on the principle of membership of a species (Hunt 2008), perhaps he has a point.

If I am to take what these authors offer on board, I will need to be aware of two issues. First is the need for research into the biopolitics of childhood to have navigational aids that are suited to the challenges and demands of the present. Second is the need to retain the capacity to make analytic distinctions without which it is so difficult to foster critical and creative insight. If Agamben is worth hearing, then the last thing childhood researchers should do is mimic questionable ruling strategies by abandoning the zoë/bios distinction and offering no replacement.

From the point of view I have been developing here, a binary frame is not the only way to navigate the field of childhood bio-politics. As long as the belief can be preserved that the world divides neatly into two chunks it retains credibility. But if Foucault is right, then the shape of childhood research owes more to the developing patterns in relationships between governments and populations than to any such 'timeless' distinctions. Fortunately, even though many interactions of multiple strategies, interests and theories have taken place in the development of the bio-political, Foucault shows that patterns have emerged. My strategy for navigation and analysis takes advantage of this.

Foucault's work suggests that the binary frame around childhood is a product of a basic concern of bio-political governance. Running in parallel with governance, sometimes criticising it, sometimes refining it are the dualistic traditions of childhood research that concern themselves with what resources in terms of capabilities and proclivities children present and the means and degree by which they can be developed, trained, refined and so on. Together, as they advanced curiosity into 'human nature', studies of child development and socialisation also informed interventions in childhoods designed to promote security for the larger and future populations. For example, Parsons (1956) considered how societies take hold of children as resources for the reproduction of values and social orders, while Piaget (1927) focussed on the sensitivity of the conditions underlying the development of 'rationality', a development that he understood on the model of other life processes. Thus, concerns for 'life' and for 'resource' have been threaded through childhood research for many years, weaving it closely into attempts to respond to political concerns.

In the twentieth century, bio-politics became yet more complicated. The German, Soviet and Chinese experiences of the twentieth century established that sophisticated states could not be relied upon to remain sensitive to the wellbeing and security of their populations.

They seemed also to possess tendencies to waste and to eliminate the lives of sections of their populations and to consider others as simple resource. In response, a Universal Declaration of Human Rights (UDHR) and, later the UNCRC were drafted and signed by most of the states of the world in the hope of checking these evident tendencies. These establish the principle that there are inalienable rights to life held by all individuals regardless of the preferences and plans of states. This was accompanied by another set of important rights – to have a 'voice' in public and political debate. The UNCRC recognises a 'voice' for children that should not be usurped by the adults, families and states that they otherwise 'belong' to. This bio-political innovation created the space for today's concerns with children's voice (Alderson 2000) and participation (Hart 1997). So far, my sketch of the intertwining development of bio-politics and childhood research suggests that as they have met, they have generated three key formations of concern, curiosity and action; Life, Voice and Resource.

The philosophers Deleuze and Guattari (1988) often use the term 'multiplicity'. It describes these formations of childhood very well. Multiplicities are gatherings of practical, political, theoretical and empirical concerns and activities that have dynamic relations with each other that allow them hold them to hold together over time. The multiplicities Life, Voice and Resource are the main features of the complex patterns that have emerged as children, states and researchers have met. As I have suggested in the case of 'voice', these patterns certainly do evolve and change, but not according to a predetermined programme. They are the emergent result of the history of interactions between states, children and research. It is entirely possible to approach them with a binary frame and to squeeze their contents into it, splitting them between the biological and the social, or life processes and life styles or zoë and bios. It has made sense to many childhood researchers of the past to do exactly this. As the following sketch of the three multiplicities will indicate, I do not propose to do so.

The multiplicity 'Life' does concern life processes, but not exclusively those that are involved in the development of individual children. Here matters of life do not lead directly to the question of human nature nor to the question of the malleability or otherwise of human resources. It also includes consideration of the trans-individual life processes that are studied in such disciplines as demography and epidemiology. Further, this multiplicity is never just about life processes but contains local and specific links to a para-biological set of concerns such as 'rights' to life and notions of a 'good' life that allows for human flourishing (Sen 2001). So the multiplicity of Life has much greater scope than the 'nature' of the binary frame. Rather than staying neatly in place as one side of an equation that when paired with the 'social' creates adult individuals, the multiplicity Life, like Resource and Voice has a 'quality of sprawl' (Murray 1999). This is because it has been continually composed of a changing set of more or less successful and productive links between events and processes that are biological, medical, legal, ethical and political in kind.

'Resource' certainly is about decisions over the value of children in the present and for the future. It is the site of arguments about what constitutes a proper use or abuse of children. It is also the site of attempts to regulate their use and treatment that are informed by the basic tension between their present value as, say, sources of labour or subjects of scientific investigation and their future value as sites of investment. But Resource is not just about the economic and the legal. Issue of children's value and use are shot through with matters of human variation such as ability and disability, gender and ethnicity that are, at least in part, embodied. Resource also sprawls into the question of children's agency that has had so much social scientific play in recent years (Oswell 2012). What resources, human or otherwise, are available to children to express or construct agency for themselves and to what extent do they consider themselves and others as resources in their own plans for the future?

The multiplicity 'Voice' is perhaps most familiar in the use of the term 'voice' to describe children's representation and participation within organisations and decision-making (Hart 1997). But it is also closely connected with life processes since it concerns 'growing up' as a passage from voiced but speechless infancy toward self-expression. Understandings of this passage have consequences in turn for the many circumstances in which children can and cannot find voice and be heard. Sometimes their immaturity is treated as a reason not to hear them. There is a further range of institutional and technological conditions which take up children's voices for various kinds of interpretation, mediation and amplification (Motzkau 2010). Attempts to capture children's autonomous and self-expressive voicings and to convert these into politically effective forms of representation have been major processes in the emergence of this multiplicity.

Mosquito teen deterrent: Life, Voice, Resource

If Agamben is right, the binary frame was constructed in performative acts of self-limitation undertaken to establish the legitimacy of rule and has gone on to act as a guide for governmental action and critical thought for millennia. Childhood is one of the sites at which it has been deployed most effectively. For Agamben, however, contemporary rule is often conducted through a blurring and effacement of this distinction that stymies critical thought about the legitimacy of power. The evidence I presented in Chapter 1 about the fading pertinence of the binary frame in a time of climate change and biotechnological advance does not follow Agamben's account, but does present a parallel problem. How are childhood researchers to find their way through a bio-political scene that does not respect binary frames and how are they to retain capacities to think critically and creatively about the rule of children's lives?

The multiplicities Life, Voice and Resource are my attempt to describe the patterns that have emerged over many years in the construction of childhood. That history provides the conditions of possibility for the application of binary frames, but does not necessitate their use. My suggestion is that these three multiplicities can assist in making connections among and drawing comparisons between key sites in the contemporary bio-politics of childhood. These connecting and comparing functions are basic to critical commentary and creative research and action. This approach certainly does uncouple childhood research from the question of 'human nature'. My view is that if answers to this question are sought they are no more likely to crop up around children than around any other grouping of humans. But, as this section suggests, it also alters the range of sites that appear to be interesting or informative.

Research that sat comfortably within the binary frame, what we might call traditional childhood social and psychological research, tried to understand the relative measure of both life processes and social processes. It asked what mechanisms allowed the two to interact. In other words, it was committed to the study of human nature through the example of the child. The 'new paradigm' work on the construction of childhood (James and Prout 1997) took aim at universalised views of childhood and in doing so emphasised the social and the political aspects of childhood over life processes. In my view, the challenge now is to find ways to address biologies, economies, regulation and childhood experience together, without returning to the 'human nature' agenda and without collapsing everything into the category of the political. I have found the following example useful to think with.

The 'Mosquito' is a product of the UK company 'Compound Security Systems' (www.compoundsecurity.co.uk). It is a device that can be mounted on the outside walls of buildings. It is capable of emitting a very high-pitched sound at an intensity that causes discomfort for anyone who can hear it. Its operation can be triggered either automatically through a movement sensor or through a human operator button push.

There is a universal developmental trend in human hearing. As humans age, the sensitivity of their hearing to higher frequencies diminishes. The consequence is that the younger people are, the more likely they are to be able to detect very high frequency sound. The Mosquito tone is pitched high enough that most people past their early twenties will be unable to hear it. For those who can, the sound is very unpleasant. This device, then, takes advantage of a universal life process – the decay of high frequency hearing – to make an unpleasant sound that can only be heard by relatively young people. The stated purpose of the device is,

> to deter youths from congregating in large groups and acting in an anti-social manner as well as causing damage to property.
>
> (www.compoundsecurity.co.uk)

Simply stated, when young people arrive at a place they are not wanted, use of the Mosquito will make their experience of that place quite unpleasant, increasing the chances that they will go elsewhere. Since the Mosquito's launch in 2006, more than 8,000 units have been sold worldwide. Customers include police forces, local authorities, retailers, car parks and security forces. The term 'anti-social' above indicates situations of actual or possible disagreement between generations over the appropriate use of space, often urban. These are conflicts, specifically between the young and property owners or authority figures.

In my terms then, the Mosquito is a prime illustration of the contemporary bio-politics of childhood. Though a universal life process is involved, it has very little, if any, connection with the grand puzzle of 'human nature'. It is instead an example of the forging of a direct connection between a life process and a political process. Crucially, it replaces any need on the part of the user of the device to communicate verbally with the young people whose conduct they would like to influence. In Agamben's terms it is an attempt to turn negotiation over the use of urban space into a site of exception that is about non-negotiable life processes rather than negotiable values.

By 2008 a campaign had been organised against the Mosquito. The 'Buzz Off' campaign (www.11million.org.uk/youth/buzz_off_campaign) drew together the energies of children's advocacy groups, UK Children's commissioners and civil liberties groups. They argued that because it potentially affected the movements of any children in the area of its deployment, it constituted a groundless constraint on rights to movement and assembly. Beyond this they also registered their disgust that children were being treated as if they were pest animals. Similar sonic devices are also used to deal with rats and mice. This response asserted children's value as human beings and strongly articulated the views and feelings of many children.

These events do not key into the question of whether and how to shape children for the future, nor into the question of 'human nature'. They involve the dissolution of apparent boundaries between life and social processes. Thus a binary frame fits poorly but, arguably, critical attention is still merited. These are just the circumstances in which I would suggest Life, Resource and Voice are good guides to the kind of questions that could be asked, the issues that need to be incorporated in arguments and to how this Mosquito example could be connected and compared with others.

Consider Resource first. Among the Mosquito customers in the UK were local government departments and retailers. If we consider their likely motivations, we can open the issue of how they related to children as resource. One aspect of the role of UK local government is to provide and maintain leisure facilities, such as swimming pools and parks. When parents see their children enjoying these facilities the benefits of paying local taxes are made tangible. So, in this sense, local governments need children as a kind of resource of political persuasion. In a similar vein, retailers often rely on children as customers. So, in this sense, children figure as an economic resource. Looking at urban space from another point of view though, it is clear that children and young people have their own ways of using buildings and facilities as resources. They are sites around which social interaction can be organised and at which children become of value to each other as friends and peers and, sometimes, sexual partners. These articulations of resource do not always result in conflict, but they can. Children's use of urban space can become inconvenient or unacceptable to adults. This resource tension provides the niche for the Mosquito as an apparently efficient way to dispose of those aspects of children's movements and conduct that cannot be recouped as resources while retaining the economically valuable aspects. The Mosquito case, then, raises a number of questions about Resource. Not all local governments or retailers have bought Mosquitos, so have others found different, perhaps more effective ways of managing intergenerational tensions of resource? What are the children the device is aimed at achieving for themselves by gathering at, say, the entrance of a convenience store? Unless their sole purpose is to intimidate other customers, it may be that they are using urban spaces to learn about each other and themselves and to establish norms.

Moving onto Voice, one key innovation of the Mosquito is that it allows for a sonic but wordless intervention. The operator who is charged with managing children as resources need never speak to them. This means that they will not have the opportunity to reply and that the operator need never engage in argument with them over the rights and wrongs of their presence. But, as the Buzz Off campaign indicates, opposition was voiced and replies were made at the relatively articulated and large scale of a national media campaign. How was this voicing made possible? Did the voices of children affected by the mosquito percolate up to the level of Children's Commissioners? If so, by what routes? Were there children who felt that the Mosquito could actually help them in their daily lives addressing a problem of peer intimidation? If so, what happened to their voices?

Given that the Mosquito depended on a universal life process of hearing loss, one kind of connection with Life is evident. The story was attractive to UK media, in part because it connected with two major public issues of the time. First, that adult's safety and lives were at risk from street attacks by groups of young people. Second, that children's health and well-being was undermined by restrictions on their use of public space for socialising and play. These concerns are part of the Life multiplicity. They weave biologically based accounts of health and development together with issues of rights to life, of autonomy and of citizenship provoking a range of questions about urban living. In what ways are children's autonomous use of space related to their quality of life? How are opportunities for such autonomy distributed socially and geographically? Which demographic clusters share sets of needs and interests in these matters and which differ?

Conclusion

Critique of the application of binary frames to childhood is well established. Prout (2005), in particular, has shown that childhood researchers are not obliged to rely on the binary frame by drawing out its associations with the ontological commitments of a now past 'Modernity'. In this chapter, I have drawn links between the binary frame and Foucault and Agamben's concerns with bio-politics. I argued that bio-political governmental strategies called the figure of 'human nature' into being as a way of organising the search for intervention in and influence over populations. At the same time, these strategies composed patterns of intervention into and knowledges about childhood that stood outside the binary frame. I called them 'multiplicities of childhood' and proposed them as a framework for examining contemporary childhood bio-politics. I offered the 'Mosquito' in illustration, and, in Chapters 4 and 5, I will consider relations between childhoods, 'mental capital' and 'philanthrocapitalism' in this way.

My aim has been to respond productively to what I see as a major tension. On the one hand, if my arguments in Chapter 1 hold good, the binary frame is clearly limited and potentially misleading. On the other hand, if we simply do away with the binary frame or declare it 'blurred', we stand not only to lose a valuable navigational resource, but, if Agamben's concerns apply, we also stand to lose critical capacity. This chapter has begun the business of developing productive alternatives that could be of use to childhood researchers facing issues of climate change and advances in the life science advances. One place to start is with the following questions, composed in the light of childhood multiplicities, that can be asked in any bio-political circumstances and will crop up throughout the book.

- How are children's voices being composed, transferred and circulated?
- How are children figured as resource and as users of resource?
- How are children's life processes engaged with in these circumstances?

3
Childhood and Bio-social Imaginations

So far I have argued that, throughout its history, childhood research has formed and reformed itself in response to changing bio-political circumstances. Individual researchers have devised strategies and approaches to find a niche in the marketplace of ideas, and thus, according to their lights, to improve children's lives. On a larger scale, and as a result of this activity, the major contours of childhood research have emerged. I have argued that while the field of bio-political action is shaped by and has shaped multiplicities of Life, Voice and Resource, childhood research has tended to navigate according to a narrower binary frame that supposes sharp distinctions between nature and nurture, and biological and social forces. The 'natural' side of the distinction is often depicted as old and slow to change, while the 'social' side is often depicted as young and fast changing. The principle appeal of the binary frame is that it promises to distinguish what can be changed about people from what, in practice, cannot, and thus to discover how to raise children so as to maximise their potential. It presents children as human futures and promises to deliver expert insight into how, in practice, to treat them as such.

Along the way I have illustrated the ability of frames like the binary frame to give order and purpose to thought and action, research and practice. I've also presented historical illustrations of the power of framing and reframing life processes and lifestyles through Agamben and Foucault. I hope to have shown that framing is a work of practical imagination. It is practical in the sense that it is undertaken in order to respond to circumstances. It is a form of imagination because, though it can be informed by factual knowledge, it is by no means thereby constrained. In this sense frames are not so much descriptions of the world as productive ways of 'performing' aspects of the world. Since childhood is a dynamic arrangement of events and forces, human expectations, curiosity, needs and imagination are central to the creation of human futures. The binary frame is only one example of such 'biosocial imaginations' (Lee and Motzkau 2012b) that have relevance for childhood. Others have been developing over the past few decades, some emerging from the traditions of the life sciences, others from the social sciences. This chapter makes a necessarily selective, comparative survey of them.

If, as this book supposes, children are creatures as well as people, then evolutionary perspectives that draw deeply on a Darwinian scientific tradition (Darwin 2003) are relevant. I will examine three perspectives of this kind: evolutionary psychology (Buss 1995), epigenetics (Bjorklund and Pellegrini 2000) and the Mismatch thesis (Gluckman and Hansen 2006). The first two are centrally concerned with what Buss (1995) calls the 'mechanisms' that underlie human behaviour and the latter, in Gluckman and Hanson's (2006) version, with the many processes that shape human diet, exercise and health. Each approach explicitly rejects a binary frame. Refusing to assume that 'nature and nurture' or 'nature and society' are different ontological realms shaped by different forces with different average rates of change, each provides its own account of the encounters between forces and the timings that shape human life. Considered from the point of view of childhood and bio-politics, each of these alternatives has its strengths and weaknesses. As I develop my own frame – 'bio-social events' (Lee and Motzkau 2012a) – my aim is, if not fully to incorporate their strengths, then at least to remain compatible with them.

The limitations of 'Blurring'

One way to describe the contemporary bio-political situation is to say that distinctions between, say, 'nature' and 'society' are being 'blurred'. If this appraisal leads us to imagine that the world we live in is itself becoming more 'blurry' then, in my view, we will have missed the point. If things look blurred to us then we just need to look with better eyes. In the social sciences this means finding more appropriate kinds of distinction than the gross two-category set that history has accustomed us to. The three multiplicities of the previous chapter are a step in that direction. But there is more to register and take into account. There is no single 'nature' but a host of 'life processes' that are as different from one another as human digestion of carbohydrate differs from plant photosynthesis which, in turn, differs from bacterial reproduction. Likewise, there is no single 'social' but a host of 'social processes' that are as different from one another as the negotiation of power and responsibility between an adult and a child differs from the emergence of industrial capitalism. Given this, we need to become discerning and mindful about the frames we choose. We need frames in which it makes sense to ask which life processes (conventionally understood) relate to which social processes (conventionally understood), over what time scales and whether and how these relationships can change. If the binary frame is 'Frame 1' then our next is 'Frame 2'.

Frame 2: Evolutionary psychology: Multiple temporalities

> ... 'nature versus nurture', 'genetic versus environmental', 'cultural versus biological,' and 'innate versus learned.' These dichotomies imply the existence of two separate classes of causes, the relative importance of which can be evaluated quantitatively. Evolutionary psychology rejects these false dichotomies.
>
> (Buss 1995: 5)

As Buss argues for using evolutionary insights to unify the discipline of psychology, he couldn't be clearer about the distance that he wanted to establish between evolutionary psychology and the binary frame. An account of the basic principles of his approach should make it clear just what sort of alternative he had in mind.

The term 'evolution' has come to mean the process by which living organisms develop and change their form and functioning over generations and in response to environmental demands. Rather than thinking of life as the creation of a deity who brought organisms into existence out of nowhere, ordaining their design or developmental parameters in an instant, the evolutionary view supposes that the processes through which species evolve always takes time, depends on many small interactions that have no guiding hand and have no end-point. Form and function always remain, in principle, open to change.

Evolutionary thinkers have their differences (Gould 2000; Dawkins 2006) but there is a clear consensus. Only three simple things are needed for the evolution of organisms to take place; reproduction, variation and selection. Reproduction does not always produce identical copies. Where reproduction is sexual, involving two parental organisms, offspring combine existing characteristics in new ways. Whether reproduction is sexual or asexual, infidelities in the complex organic processes of reproduction also inject variation into the new generation. Once variation and diversity are in place, the key question is how they affect the ability of the new generation itself to reproduce – its reproductive fitness. If a given variation gives one individual an advantage over another when it comes to finding food, avoiding injury and fatal hazards, or attracting a mate, then there is a good chance that, in following generations, that variation will become increasingly common in the population as a whole. Selective pressures exerted on a diverse population of organisms over long periods of time can have cumulative effects, resulting in the emergence of new biochemical processes, cellular structures, organs and even, as Darwin (2003) argued, species.

Evolutionary psychology grounds itself in this view and supposes that if evolutionary processes can produce organs as complex as eyes and brains, then it could also produce psychological mechanisms that underlie human behaviour. The central hypothesis offered by evolutionary psychology is that such evolved behavioural mechanisms exist and that many, if not all, have discrete or 'modular' functions, rather like bodily organs. As Buss has it, this central hypothesis is too general to be directly tested. Thus, the bulk of evolutionary psychologists' work lies in deriving subsidiary hypotheses from the central one and then testing those hypotheses against empirical evidence regarding human behaviour. As subsidiary hypotheses are tested, adapted and made a better fit for empirical evidence, Buss argues, so the central hypothesis gains in credibility.

If Buss were strongly influenced by two-category thinking he might develop his argument as follows. We know that certain human behaviours were established in the past through processes of variation and selection. We cannot simply undo our evolutionary history because, first, it is buried deep in our bodies and is thus inaccessible to both persuasion and force and, second, because as individuals, groups and societies we simply do not operate on the long time-scales needed to alter that which has evolved. In other words, many of our behaviours, especially those closely associated with reproduction, are driven by unchanging instinct. For Buss to take this direction he simply needs to allow that there are just two relevant temporalities at work in human behaviour: the deep past and the present day. This would reproduce the characteristic set of tensions inherent in the binary frame: two different kinds of force, one old, slow and biological, and the other contemporary, fast and social, that come into conflict whenever humans try to shape or control themselves or one another.

Buss explicitly rejects this view however. For him, understanding human behaviour requires attention to a wider range of temporalities. He examines three kinds of time and context; the historical selective context, the ontogenetic context and immediate situational inputs. The historical selective context is what might most readily come to mind when thinking about the evolution of human behaviour. Consideration of this context raises questions about the kind of environments that humans evolved in and the selection pressures that these environments brought to bear on our ancestors in all their physiological and behavioural diversity. In other words what did ancestral humans have to do in order to live long enough to have and to raise children? Crucially, we should not imagine this as the time during which fixed instinctive behavioural patterns – an homogeneous human nature – were laid down deep in the human psyche. This is because, for Buss, there is no good reason to suppose that the environments humans evolved in were themselves stable and fixed. To him, it seems likely that adaptability to changes in human and non-human environment would have been key to survival.

As he turns to describe ontogenetic context, Buss models this adaptability as a repertoire of different behavioural strategies. The term 'ontogenetic context' captures that segment of time we know as 'growing up', as psychological development or as socialisation. While the historical selective context furnishes a repertoire of behavioural strategies, Buss supposes that details of the ontogenetic context can shift an individual towards one strategy and away from another. To illustrate the point he reports evidence that fathers' absence during childhood is correlated with individuals having higher number of sexual partners in later life, while the presence of an 'investing' father 'shunts individuals toward a more monogamous mating strategy' (Buss 1995: 11).

Third, Buss draws out attention to what he calls 'immediate situational inputs'. Just as the life processes that can thicken and callous the skin on the sole of your foot are only activated where that skin is exposed to friction, so evolved psychological mechanisms only become operant under certain conditions. Buss offers the example of 'discriminative parental solicitude'. Parents typically give their children a lot of care and attention, but that 'solicitude' is reduced in certain circumstances; where a parent is not a biological parent or where the biological relation is in doubt; where an infant is deformed or seriously ill; and, where parents' poverty sets limits on the number of children they can successfully raise. Indeed, according to Buss, these are the circumstances in which adult carers, regrettably, become capable of killing infants.

For Buss, the apparent switch between profound love and indifference or violence that these findings speak to is best explained in terms of reproductive competition between adults and competition for resources between children, aspects of human behaviour that only come to light in certain circumstances. Discriminative parental solicitude is good example of the distance between evolutionary psychology and two-category thinking. Buss is not presenting us with the idea that, in extreme circumstances, our beastly nature is revealed to us in compulsive and instinctive infanticide. Parental indifference and violence are not signs of a primitive inheritance overwhelming civilised social mores, codes and regulations. Rather, both the capacity for intense love and that for indifference represent evolved behavioural strategies. Which one is expressed depends on circumstances.

My intention in presenting evolutionary psychology here is not to convince you that it is correct. The point is not that Buss has all the answers, indeed, as broad as his canvas is, in the rather broader terms of reference of this book, its exclusive focus on individual human behaviour is an unnecessary limitation. Rather, Buss presents a way of posing questions that breaks the binary frame. Crucially for Buss, behavioural tendencies created in the human evolutionary past should not be understood as a dead hand automatically determining the present, or as presenting behavioural decrees from within a space inaccessible to the vagaries of present circumstance. There is no genetic determinism here. Rather, on his view, the present bundles pasts together in the sense that present individual behaviour is a result of the activation in the present of embodied expectations formed yesterday, in childhood and in the last 200,000 years. This is a temporal imagination just as much as a bio-social imagination. It asks us to see and to suppose that everything happens in the present as different temporalities are drawn together in interaction with one another.

Buss imagines human behaviour as the emergent result of an interiority of temporal relation in which embodied expectations meet circumstances. Such encounters happen countless times every day and those expectations continue to shift. The way evolutionary thinking analytically distinguishes between different temporalities and then figures the present as a mixture of these temporalities has led one commentator to describe the emergence of biological form and behaviour as 'untimely' (Grosz 2004). On this view, life processes are not to be understood on the model of an organism carrying out the instructions written in its genome, as if reading from a script, but are always innovative and open-ended encounters between multiple forces.

This evolutionary approach, however, does not suggest that all of our embodied expectations can change in an instant. Such a claim would collapse the multiplicity of times that Buss describes into a formless and unconstrained present. An organism with no expectations about its environment is not likely to prosper. After Buss' careful expansion of the list of temporalities of the bio-social, the image of the 'natural' and the 'social' in constant tension with one another looks less than convincing as a model for bio-social phenomena, including childhood. But considerable scope remains within this evolutionary paradigm for 'bad timings' to occur, occasions at which embodied expectations and circumstances are at odds with one another in ways that can be experienced as 'problems'. I will shortly discuss the identification of such 'mismatch' in the contexts of child behaviour and health. For the moment I should acknowledge some of limitations of evolutionary psychology as a form of bio-social imagination.

First, empirical evidence of selective pressures in human evolution is only as strong as our accounts of human pre-history are complete and, though such accounts are impressively detailed (e.g. Finlayson 2009), many questions remain. This leaves evolutionary psychology open to the claim that its claims cannot be falsified (Richardson 2009). Second, evolutionary psychology presents an attractive metaphor for 'shunting' between different behavioural strategies, but offers no material account of the mechanism involved. Third, at times it makes statements about human behaviour, particularly gendered behaviour, that meet with political objection (ibid 2009) and in these encounters is often left restating its founding assumptions about the relation between observed behaviour and selection pressures of the evolutionary past rather than demonstrating them to be correct. But if we pay attention to the way it stages the bio-social, it becomes clear that evolutionary psychology ignores the boundaries of the binary frame so as to register relatively more complexity and dimensions of variability. Because it does not ask us to see the world as 'blurred', it remains articulate about how behaviour can be studied.

In this sense it can be seen as a relatively new variety of bio-social imagination that sits alongside and in alternative to the binary frame that has, according to Agamben, been at work for millennia. Its key feature from my point of view is that it supposes that, even though the temporalities of human conduct are complex, even though a straight split between the ancient and the immediate cannot be supported, it remains possible to draw distinctions within that complexity so that it can be analysed.

Frame 3: Child development and epigenetics: Continuities of life process and lifestyle

The sociologist Qvortrup (1994) introduced the argument that children should be understood as 'human beings' rather than 'human becomings'. This was a way of saying that what children do outwith adults agendas and interests is of sociological and practical importance. Bjorklund and Pellegrini (2000) develop a parallel view in the context of evolutionary psychology. Though Buss has a lot to say about the behaviour of adults around children, children's own behaviour is often given less emphasis. Bjorklund and Pellegrini suggest that this is because children, who are not involved in sexual reproduction, are seen by some evolutionary psychologists as further from the evolutionary cutting edge of adaptation to selection pressures. They then point out, however, that for an individual to reproduce, childhood itself must be survived and, thus, that some patterns in children's behaviour and psychological capability might be best understood as evolutionary adaptations that, under many circumstances, increase rates of survival to adulthood.

Critics of developmental approaches to childhood have often problematised the ideas that development is linear, that it has an 'end', and that a 'normal' path of development can be discovered that is desirable because it is 'natural' (Burman 2007). Such critics alert us to the close relation between such commitments and the governmental rationales in which children figure as human futures – resources for state security. Once again, Bjorklund and Pellegrini, for their own distinctive reasons, develop a parallel view. They point out that much developmental research is concerned with the early identification of traits that may show up in adulthood with an eye on the encouragement and/or control of these traits. This early identification of traits to be encouraged and discouraged is a feature of many applications of the binary frame. In contrast, they try to broaden this agenda, seeking to account for observed patterns of children's behaviour and capabilities in terms of their adaptation to the task of survival till reproductive maturity. In other words, they want us to imagine the ongoing present of childhood experience, not only in familiar terms of social interaction, but, simultaneously, as a site of biological adaptation and change. For them 'ontogenetic adaptations' that are characteristic of childhood are not simply incomplete versions of adult characteristics, but have roles specific to the challenges of childhood. They offer a number of examples.

They tentatively define 'play' as behaviour that has no apparent function or for which engaging in the behaviour itself is more important that any results it brings. They consider the future-oriented view that children play a great deal in preparation for adulthood and offer the complementary view that playing is significant in the 'present'. In the course of play, relationships between children and other people and children and objects are transformed in ways that may predict future behaviours, but need not. For example, they describe children's object play in terms of the opportunities it makes for finding culturally innovative ways of doing things and using things. Here play is not just preparatory but is potentially transformative. They describe 'rough and tumble' play both as means by which leadership within peer groups can be established and challenged and as a means for promoting skeletal and muscular development. Again, these activities may be seen as preparatory but they can also play into the degree to which children are able to survive to adulthood. Even though play always has costs in terms of the energy it uses, its adaptive benefits would seem to outweigh those costs.

Considered from a standard developmental point of view, children's cognitive immaturity is no mystery. Adult human cognitive abilities are complex and hard to achieve, so, starting from a naturally low and limited set of capabilities, children simply have to follow a long road to reach them. Bjorklund and Pellegrini present children's cognitive immaturity as more of a puzzle by turning that line of thinking on its head. Supposing that children could develop high cognitive abilities at a very young age – and some prodigies certainly do – and supposing that this early development would offer selective advantage, what processes could have selected for the forms of cognitive immaturity that are so widely observed? Take young children's limited ability to reflect on and accurately assess their own conduct. Rather than seek to understand this by contrast with supposed adult capability and see this characteristic as a deficit, Bjorklund and Pellegrini suggest that it is a kind of negative capability. Those less able to understand themselves as 'incompetent' are more likely to be resilient to failure. Similarly, they suggest that young children's limited working memory capacity may assist language acquisition. Limited working memory acts like a filter on the experience of language, effectively forcing children to analyse speech into its component parts.

Bjorklund and Pellegrini reject any notion that children's development can be analysed in terms of 'natural' and 'social' components, where the former are fixed and determining and the latter struggle for control. For them, just as life processes inform sociality – for example by making language possible – biological structures and functions, a set of embodied expectations, remain open to change throughout life in the light of experience. They take an 'epigenetic' approach to relations between organism and environment. Genetic activity, the conversion of DNA into RNA and the subsequent construction of proteins at the cellular level certainly shapes organs and bodies and these shapes in turn enable certain functions and activities. But the organism never stops taking its environment into account. This is because the arrows of influence go in both directions. As they put it:

> ... genes code for the production of protein molecules, which in turn determine the formation of structures, such as muscle or nerve cells. But activity of these and surrounding cells can serve to turn on or off a particular gene, thereby causing the commencement or cessation of genetic activity .... From this viewpoint, there are no simple genetic or experiential causes of behaviour; all development is the product of epigenesis, with complex interactions occurring amongst multiple levels.
>
> (Bjorklund and Pellegrini 2000: 1691)

Like Buss (1995), Bjorklund and Pellegrini distance themselves from a binary frame. What I find inspiring in their approach is that they do not then conclude that everything is 'blurry' or beyond analysis. They fend off the simplicities of two-category thinking by making different kinds of distinction – between the whole organism and its environment certainly, but also between DNA, RNA, proteins and cells.

The methodological lesson I draw from this is that to avoid a binary frame while remaining articulate about the bio-social means developing new, more incisive distinctions. I am strongly influenced here by the physicist Barad (2007) whose concept of 'agentic cuts' nicely expresses the mindful approach to the drawing of distinctions and the creation and selection of frames that I would advocate here. She is aware that in her discipline, frames and practical imagination are in constant play as efforts are made to generate and interpret data and theoretical statements. Her remarkable insight is that this mixture of human effort and desire within attempts to determine constraints and tendencies in physical systems need not result in a blur of indistinction between subjective and objective accounts. Rather, the fact that practical imagination is an inalienable part of human activity and understanding first requires our acknowledgement and, second, should be taken as a spur to the deliberate use of reframing devices in scientific research. I will return to this theme in later chapters on climate change and sustainability.

In Bjorklund and Pellegrini's case the idea of epigenesis becomes a wellspring of research questions. They take it for granted that the arrow of influence goes in both directions, but do not suppose it does so without limit or distinction. No doubt there is 'plasticity' and flexibility in organisms' developmental responses to environments, but exactly what degrees and specific kinds of change they are capable of, at what speeds and in response to what experiences remain to be investigated. A couple of examples of plasticity and its limits should help here.

Bjorklund and Pellegrini give some consideration to the question of why, if the expression of the human genome has flex, there are any behaviours that can be considered species typical. Surely, if there are universal or near universal human behaviours, this is evidence for behavioural genetic determination? Their answer is that species typical behaviours come about because species typical genomes encounter species typical environments. In other words, humans tend to share similar environments both in terms of common material challenges to survival and in terms of the fact that we grow up surrounded by other humans and all their demands and complexity. This match-up reliably results in similar outcomes. Yet there is evidence that presenting an organism with a species a-typical environment can subtly alter resulting behaviours and capabilities. One example is the unusual early experiences of pre-term infants who are exposed to species a-typical environments of the intensive care ward. Als (1995) argues that the pattern of impairments (lowered IQ, attention deficit, eye-hand coordination difficulties) and accelerated or enhanced abilities in other areas, such as mathematics, that characterise such children as they grow are an example of plasticity in response to stimulation outside the species-typical range.

This evolutionary, epigenetic approach does not present a picture of inevitable, smooth coordination of organism and environment. There are limits to the flexibility and adaptation that individuals are capable of and the nature of these limits vary within populations. For example, some children (and adults) exhibit relatively high levels of impulsivity, rapid environmental scanning and hyperactivity. Some children with these characteristics find school environments that require high levels of self-control, focus and physical stillness difficult to manage. Bjorklund and Pellegrini suggest that Attention Deficit Hyperactivity Disorder is a result of 'mismatch' between a behavioural repertoire that was highly adaptive in human prehistory and the environments many of us now inhabit. When foraging and hunting, exposed to attack from predators and in competition with other animals for food, impulsivity, rapid scanning and hyperactivity could all be useful.

From a binary frame point of view, this mismatch might look like a good example of a tension between old and slow nature and young, fast society. From within the evolutionary, epigenetic perspective, however, this example is just one island of obvious tension set in a vast sea of far more complex and variable relations. Though two-categories might be enough to give a superficial description of one set of circumstances, perhaps those of an individual child with an ADHD diagnosis, understanding them in any depth and designing appropriate responses will require a lot more distinctions. Crucially, approaching the 'nature' of children as fundamentally composed of many variable relations and, as a product of evolutionary processes, as inherently diverse means that instead of seeing ADHD as evidence of a departure from some natural developmental norm, as a deficit or a disorder, we can see it as a piece of human diversity.

Commentary

I will shortly pursue this notion of 'mismatch' through issues of health, diet and exercise as addressed by Gluckman and Hansen (2006). Before I do so, I'd like to restate some key points that will help me develop my own approach to remaining articulate in the face of untimely bio-social complexity, while avoiding two-category thinking. In presenting these evolutionary approaches, it is not my intention to convince you that they are true, or that they have all the answers. Rather I present them as examples of how fresh frames can be developed to ward off the binary frame when addressing childhood as a bio-social phenomenon.

The evolutionary approaches I have examined display little inter-est in nature/nurture, nature/culture or natural/social distinctions. Rather, their founding distinction seems to be between 'adapting unit' and 'environment'. Unlike those of the binary frame, this dis-tinction is not used to define the contents of the world prior to investigation – nature, slow and old versus social, fast and young. Rather it is a heuristic distinction, an initial articulation that is designed to generate questions. As a distinction it certainly has two parts, but it could not be described as dualistic because it actually forms the basis of a multiplicity of further distinctions. Sometimes the adapting unit is a whole organism. Sometimes it is the nucleus of a cell adapting its protein manufacture in response to cues from the organism. What counts as environment is just as variable. For a cell nucleus, the rest of the cell comprises part of its environment. For a whole organism its environment is composed of the many opportunities and constraints on its survival and reproduction. Working our way up to whole human organisms, the environment also contains other humans and all the threats and opportunities they present. Try to analyse this with just two categories and the best you will end up with is the impression that the bio-social is blurry.

The further investigation proceeds, however, the clearer it becomes that in children's behaviour, as in adults, we are seeing the results of multiple events and circumstances of concord and discord among a huge range of distinct processes each of which has its own range of flexibilities and paces of adaptation. In the notion of 'mismatch' these approaches certainly make a place for those circumstances in which life processes and social processes are in tension. These circumstances might seem to confirm the pertinence of a binary frame, but an evolutionary and epigenetic view would place such islands of 'mismatch' in a much wider context of relationships and possibilities. From this point of view, the binary frame is fixated on a rather small range of questions those that are close to the heart, for example, of a pedagogue or an educational psychologist who share in the governmental work of producing children who can fit whatever image of their futures is currently preferred.

It was thinking about the way that evolution and epigenetics sustain articulacy about the bio-social even as they ward off a binary frame that first led me to wonder what parallel heuristic distinctions could help childhood studies to adapt to a world of fresh bio-social and bio-political complexity. In the binary frame, the child is imagined as the meeting place of two great sets of forces each of which is united by a common temporality – nature old and slow, society young and fast. Seeing life scientists reject that frame takes us one step closer to 'biosocial events' as I imagine them. Even as we reject the twin monoliths of 'nature' and 'society', we may yet wish to retain a heuristic distinction between 'life processes' and 'social processes'. To draw this distinction is not to commit oneself to the idea that such processes are always distinct, but to allow oneself to remain articulate in raising empirical questions about bio-social relations. As I indicated above 'life processes' are highly diverse, as are 'social processes'. Relations between specific processes are also highly diverse. Digestion of my breakfast, for example, has implications for my blood sugar level and thus my ability to concentrate as I write this text. Similarly, my negotiation of space and time to work on this text is closely articulated with wider processes of change in the UK university sector. As should become clear, the value of the heuristic distinction between 'life processes' and 'social processes' is to draw our attention to those occasions when the nature of relations between a specific life process and a specific social process change. It should also extend the kinds of issues we are able to research beyond the issues of the explanation of individual behaviour that define Buss (1995) and Bjorklund and Pellegrini (2000).

Frame 3: Mismatch thesis

> We don't know whether the reader experiences the same sinking feeling as the authors when the phrase 'nature or nurture' come up. We regard it as an artificial and unhelpful dichotomy.
>
> (Gluckman and Hanson 2006:28)

As the above quote indicates these authors share much in their view of the binary frame with those above. For them, human beings, like any other creature, have a broad adaptive capacity. They share the view that our life processes are the outcome of epigenetic processes in which the 'outside' of environments and the 'inside' of cellular reproduction are in fact always continuous with one another. For any processes that develop within this continuity, there are points at which either no further adaptation is possible, or at which the outcomes of adaptive responses are perceived by many, if not all, as undesirable. These localised and temporary outcomes taking place in a wider field of interactions are what they describe as cases of 'mismatch'. A mismatch in their terms is not a conflict between an ancient and unchanging norm of human biology laid down in the evolutionary past and an unusual modern change of lifestyle. Rather it is the present-day result of interactions between human adaptive capacities and lifestyles that have been ongoing throughout human history.

Among other things, their book aims to account for the emergence of the so-called 'lifestyle diseases' including cardio-vascular disease, obesity and type 2 diabetes in terms of mismatch between human adaptive capacities and lifestyles. They are clear that human height, weight and health all vary historically with diet and lifestyle. There is, for them, no naturally defined normality in human functioning waiting to be recovered. Rather, 'lifestyle diseases' represent the adaptation of human life processes to circumstances in which highly calorific foods are widely available and the amount of exercise involved in daily life is, in historical terms, relatively low. They do take the view that the costs of these adaptations, in terms of increased suffering and reduced longevity, are so high that it is reasonable to see them as undesirable.

Where the epigenetic concerns of Frame 3 were focussed on questions of individual behaviour, Gluckman and Hanson take a comparable approach to matters of human health and well-being. Rather than presenting us with interactions between monolithic binary categories, their accounts are full of detailed appreciation of specific relations between particular life processes and particular social processes. In doing so they offer accounts of what have been described elsewhere as 'local biologies' (Lock and Kaufert 2001). For example, they point out that the timing of menarche varies geographically and historically with variations in diet. Where food supplies become more reliable and where public health measures reduce infectious disease, menarche tends to begin at earlier chronological ages. So, for them, two sets of different social processes have affected one life process. Taking the issue further and considering the further consequences of early physical maturity for girls living in highly complex and sexualised societies, they then discuss these events in terms of a 'mismatch' of modern puberty.

A similar focus on relations between specific processes can be found in their coverage of the issue of relatively high modern rates of metabolic syndrome in populations of the Indian subcontinent. Where some have taken the view that this probably reflects a genetic heritage, Gluckman and Hanson see the possibility of a highly specific set of relations between lifestyles and life processes. Where women have had subordinate social roles involving a lot of physical labour and reduced nutrition there will be a tendency for their children to be born at relatively low weight. Poor foetal development reduces these children's ability to adapt to the rich diets that accompany modernisation. The outcome is a higher rate of metabolic disease in Indian populations than we might expect, judging from UK or US experience.

Frame 4: Bio-social events

So far, I have shown how today's behavioural and medical scientists can now draw on understandings of evolution and epigenetics to frame issues of childhood and development in such a way as to avoid the binary frame. The principle advantage this offers is a keen awareness that any general category of 'life processes' or of 'social processes' contains an enormous variety of distinct processes each of which has its own susceptibility and capacity to interact with others. There is also a theme of interaction between multiple time scales that plays out over distinctive durations. It is this sensitivity to difference and variety that enables the various non-binary frames we have considered so far to generate hypotheses and detailed explanatory accounts of observed outcomes.

As I present my final frame, it should become clear that an approach is available from the perspective of the social sciences that is deeply symmetrical with these other new frames. Just as behavioural and medical scientists have picked up the theme of diversity in their own ways, offering insight variously into individual behavioural and collective health outcomes, so the 'biosocial event' (Lee and Motzkau 2012a) frame I introduce here offers insight into relations between the effectiveness of biological technologies, and the ethics and power relations involved in their use. As a form of biosocial imagination, this frame is designed to draw attention towards the emergence of novel connections between specific life processes and specific social processes. Since some, but not all social processes are the result of deliberate individual or collective human action, this frame is concerned equally with deliberate technological attempts to create new connections and with connections that emerge without deliberate human action. The best way to introduce the frame is with an example.

The bacterium *Neisseria Meningitidis* (NM) is often described as a 'commensal' species. That means that it normally lives within human bodies without causing any disease. In the case of NM, it is estimated that 10–30 per cent of US adolescents and young adults are carrying

NM in their noses and throats at any one time without coming to harm (Schaffner et al. 2004). Commensal bacteria do not simply sit within human bodies but are involved in lively interaction with human cells and life processes. Human cells, especially those of the immune system, engage actively with them. Most of the time, these interactions have consequences on such a small scale that it can make sense to see life processes and social processes, in line with the binary frame, as belonging to separate, mutually irrelevant ontological categories. A binary bio-social imagination can therefore be sustainable despite the fact that there is a clear physical continuity between NM bacteria, human bodies and social interaction.

Unfortunately, commensal relations between NM and human tissues do not always apply. Sometimes, depending on a wide variety of factors and processes, NM bacteria multiply so rapidly among human cells that they become an infection of human tissue. When the tissues are inflamed and damaged, NM bacteria can enter the bloodstream, thereby leaving their commensal position in the nose and throat to travel within the body. If they reach the tissues called 'meninges' that surround the brain they can cause inflammation, putting pressure on the brain. The bloodstream can also distribute NM through vital organs and limbs. In other words the result of the shift away from commensal relations in the case of NM can be the life-threatening disease known as 'meningitis'. When this happens, it no longer makes sense to use a binary frame, because the interactions between NM and human cells are producing outcomes that register at the level of the whole sick individual. This change is registered both in terms of their changing biological functioning and in the mobilisation of whatever cultural categories of health and wellbeing are used by the people around them and in negotiations of care that can involve family, community networks and medical expertise according to specific circumstances. Considered as a bio-social event, the emergence of an individual case of meningitis is an excession of the binary frame that ties the progression of cellular and organic events closely together with matters of ethics, power and responsibility. This tie is an example of the kind of bio-social novelty that the 'biosocial event' frame is particularly sensitive to.

This specific bio-social event is most likely to involve children and young people of particular ages. Infants and toddlers are more susceptible to meningitis than adults. In part this is because their immune systems have not had time to adapt themselves to the presence of NM. Individuals in late adolescence are particularly susceptible to meningitis wherever attendance at college and university is the norm. At this stage of the social process of education, youngsters from different geographical areas meet and socialise. This maximises opportunities for individuals to meet strains of NM that their immune systems are not yet adapted to. If this example is a good guide, then, we should expect specific bio-social events to be distributed among and shaped by the multiplicities of childhood discussed in Chapter 2. It is clear, for example, that one aspect of the deployment of young people as educational resource is a change in the profile of a specific disease. Likewise, issues of voice are of great significance in the well-being of young children who are developing meningitis. They may be unable effectively to articulate their discomfort and so the danger is that their condition may go unnoticed or be misunderstood by their carer. Brain damage and blood poisoning can be very rapid in cases of meningococcal meningitis so time is of the essence. Thus bio-social events can constitute new urgencies, limits or opportunities for intervention within the multiplicities that constitute children and childhood as tractable human projects.

So far I have used the example of an individual case of meningitis to illustrate the 'biosocial event' frame, to highlight the relations it sees between disease process, responsibility and power and to suggest connections that it has with multiplicities of Life, Voice and Resource. A bio-social event is a meeting of one or more life processes and one or more social processes to create a new relationship of mutual relevance between the two. In my terms, then, the innovative connection forged between age-related hearing loss and the control of urban space by the Mosquito device enacted a bio-social event. A life process was deliberately linked with a social process to encourage a certain kind of social order. This attempt to gain traction on children as resource was then met by considerable articulate resistance. Whether a bio-social event is deliberate or not, whether it affects disease or generational relationships, the bio-social event frame can be used to gain analytic purchase on it.

NM is not only involved in many individual cases of meningitis. There are also regular epidemics of meningitis that affect people, especially children in a specific region of sub-Saharan Africa. The 'meningitis belt' stretches from Senegal in the West to Ethiopia in the East. Epidemics within the belt occur in cycles of between 8 and 14 years (Moore 1992). Some of the conditions that underlie this regularity would be conventionally described as life processes and some a social processes. These include geographical distributions of different NM variants, patterns of immunity to these variants, seasonally dry and dusty air between December and June, variations in quality of diet, endemic parasite and upper respiratory tract infections and social cycles of population movement associated with pilgrimage and seasonal markets. When these factors meet in the appropriate combination the result is the bio-social event of a meningitis epidemic. I'll now consider an attempt to engineer a bio-social event using biological technology.

Pfizer is a major pharmaceutical company. It makes and markets chemicals, some of which are intended to prevent or treat disease in humans. In my terms it is in the business of producing bio-social events on a reliable and repeatable basis. As part of its normal business of drug development it has to test its potential products. The company has recently concluded a legal dispute with the Nigerian Government that arose as a result of such a test, lasted a decade and costs Pfizer £50,000,000 in an out of court settlement. Paying this money does not represent an admission of wrongdoing on Pfizer's part.

In 1996 an epidemic of meningitis emerged in the city of Kano, Nigeria killing at least 11,000 people. There were many children in need of treatment. The charitable organisation Médecins Sans Frontiéres (MSF) were distributing an established antibiotic treatment against meningitis with the brand name 'Rocephin' (Annas 2009). A Pfizer team arrived ready to distribute a new antibiotic treatment called 'Trovan'. In line with normal practice, one of the reasons for the Pfizer team's presence was to test the effectiveness of the new drug in an active comparator study. They selected 200 sick children to take part in their study giving half of them Rocephin and half of them Trovan. The purpose of the study was to compare the effectiveness of the new drug with the established drug. An active comparator design was chosen in line with ethical considerations and steps had already been taken to establish that Trovan was safe for children. Out of the total of 200 children, 11 subsequently died, 5 of whom had been given Trovan treatment and 6 of whom had been given Rocephin. The case hinged on whether parents claim that they had not given informed consent to their children's involvement could be substantiated. Pfizer maintains that informed consent was obtained.

Consider the epidemic from the point of view of the frames that might be applied to understanding and responding to it. A binary frame may appear to be a good fit for the circumstances. After all, here we have a deadly disease that looks a lot like a natural phenomenon and, thanks to the social processes that have led to the development of pharmaceuticals, we have a human-designed challenge to that disease. Here the natural and social are placed in conflict with one another. This is a rich imaginative seam, pitching heroic medics against uncaring life processes. The alarming pace of a meningitis epidemic provides just the sort of conditions in which a rapidly organised clash between man and nature would seem to be a sensible response. A clear narrative of causal responsibility that blames NM calls for resolution by the chemical destruction of NM. Even though on my view this epidemic was a far more complex bio-social phenomenon than this picture would allow, it can certainly seem appropriate to simplify things by blaming NM when, for example, a child's life hangs in the balance.

However, an existing proven treatment was already available in Kano. What could justify Pfizer's emergency arrangement of its active comparator study? Clearly, taking the opportunity to test Trovan was consistent with Pfizer's mission to develop new drugs that could, arguably, be of great benefit to children in the future. There was always the chance that Trovan would prove more effective than Rocephin under local circumstances. As long as we are aware that such epidemics occur with regularity and are occasioned by combinations of social and life processes however, our options for response are not limited to the development of new drugs. The general health of the population, their resilience to dry dusty weather, people's movements and the way health systems are organised all offer different angles of intervention. Poverty reduction and the organisation of health care fell outside Pfizer's remit in 1996. But biomedical ways to manage NM could have been found in less urgent circumstances. For example, since local populations have developed a mistrust of pharmaceutical corporations over years of experience (Annas 2009) Pfizer might have concentrated effort on changing those relations for the better, increased their knowledge of local conditions and made £50,000,000 settlements less likely. Further, as Hope (2008) argues, biotechnology communities in developing countries are often hampered by the intellectual property rights with which pharmaceutical companies defend their interests. It might make sense to establish cooperative relationships between wealthy companies and majority world researchers. I'll be returning to these issues in Chapter 5.

'Biosocial Events' in summary

To summarise my account of this frame, it uses a contrast between life processes and social processes as a device to expose the differences within these categories and to develop sensitivity to the many moments at which different processes can come into relevance for each other. Applying this to the politics and ethics of health care multiplies the range of questions about and options both for responding to bio-social events and for seeking to precipitate them. Like the other non-binary frames I have examined in this chapter, the bio-social event frame is committed to diversity and to adaptation. It adds special attention to ethical and political dimensions of bio-social interventions. It draws attention to the way different frames of bio-social imagination can shape interventions and outcomes. Further, the bio-social events it depicts are often distributed within the spaces shaped by Life, Voice and Resource. The Kano parents raised questions about the way their voices were enabled or disenabled. Pfizer risked appearing to use children as a resource in their commercial strategies and the special value placed on children's lives arguably adds to the credibility of the narrative of urgency that encourages us to see nature and society in conflict. Arguably, it was this binary form of bio-social imagination that shaped Pfizer's actions as they rapidly organised their response.

Conclusion

In this chapter I have identified the binary frame as one form of practical bio-social imagination among others. These imaginations matter because how relations and possibilities for relation between life processes and lifestyles are conceived of has major effects on the responses to difficulties that emerge. Taking the example of a child diagnosed with ADHD. See this as the result of a determining genetic heritage and the child has a disorder that needs treatment. See it instead as the result of as yet undetermined interactions between life processes of many and various kinds and social processes of many and various kinds and a more experimental approach to securing the child's happiness and well-being might emerge. This could start off from the view that ADHD is just an instance of human diversity. In practice, around ADHD, both of these forms of imagination are often operant, themselves interacting in complex ways. Similarly, a bio-social imagination that presents expert humans in conflict with uncaring and rampant pathogens encourages certain kinds of response to disease, while a bio-social imagination that accepted a wider range of bio-social events as contributors to epidemic could offer a wide range of potentially effective responses.

My purpose in presenting a diversity of forms of bio-social imagination is to illustrate the links between the way relations between life processes and lifestyles are imagined with the response people make to difficulties. As yet it is not entirely clear how some of the frames I have considered here will improve or otherwise impact children's lives. For me, however, the important thing to notice is that choice about how to frame and thus respond to our circumstances not only persists with developments in the life sciences, but is today becoming of greater importance (Barad 2007; Wagner 2009). I will develop this thought and its implications further when I turn to the topic of climate change in Chapters 6 and 7.

4
Childhood and Mental Capital

In recent years, a form of bio-social imagination has emerged that frames children and the life processes underlying human mental abilities and states and seeks to order their relations, assigning each a clear role in a context of global economic competition. Within this frame, it is asserted that children's minds are a resource, that expertise exists on how best to deploy and to foster that resource and that states need to encourage the application of this expertise to secure future prosperity and well-being for their populations. Although this frame was articulated with particular clarity in the United Kingdom in the last decade, varieties of it inform an internationally relevant literature (Cooper et al. 2009).

Some of the discussion surrounding this mental capital frame has been organised by the concept 'cognitive enhancement'. This is often understood to mean deliberate intervention in life processes with the intention of reliably increasing powers of concentration, memory or reasoning beyond a present 'normal' level (Office of Science and Technology 2005). As some point out, the education that takes place in schools and universities could count as a major and primary form of 'cognitive enhancement' (Bostrom and Sandberg 2009), but the term is mainly used by researchers and commentators to describe what they take to be novel approaches. These are often biotechnological rather than institutional in character, promising reliable and direct effects at the level of individual cognitive performance. Sometimes discussion focuses on genetic manipulation, sometimes on human/computer interfaces, but most often on pharmaceuticals and other substances that are understood to have the potential to alter brain chemistry.

There is a wide and growing literature about pharmaceutical 'cognitive enhancement' in which two sets of issues are of particular concern. First, in the case of any candidate cognitive enhancer there is a series of questions about effectiveness: Does use of the substance actually improve cognitive performance? Is the improvement the result of real biochemical action or is it merely a placebo effect? Does it have unwanted side effects for the user? The idea that a substance might be developed into an enhancer sometimes starts with a trial of a substance within clinical populations who have a specific cognitive deficit or other medical issue. If Modafinil can regulate otherwise pathological sleep patterns, for example, it might also offer wakefulness benefits to healthy individuals (Greely et al. 2008). Sometimes it starts with the informal use of drugs, such as Ritalin, prescribed for other purposes but used in attempts to sharpen exam performance. In these settings, standard mechanisms of drug testing and development are used to address effectiveness issues. As yet, evidence for the effectiveness of candidate enhancers is scarce and mixed (Williams and Martin 2009).

The second set of issues concerns the impact and regulation of cognitive enhancers considered from an ethical and political point of view. In this literature (Harris 2007; Sandel 2004) it is often supposed that there already are or soon will be forms of cognitive enhancement that are or will be reliably effective at the level of individual performance. Often this supposition is then used as the basis for ethical and political speculation about near future events. The basic questions posed are something like this: 'If it were possible to take a pill to boost exam performance, what might the consequences for individuals and society be and are they desirable?' This genre of ethical inquiry is one way to relate values to biotechnological change. It asks us to imagine that the change has already happened and then challenges us to consider how we actually feel about it. Here, sets of values are considered in the light of facts that have not yet arisen. For example, given that enhancers might be expensive, they might put downward pressure on social mobility by helping the wealthy to succeed in public examinations (Sandel 2004). So the kind of question considered here is how enhancers relate to equality of opportunity, a hegemonic value position adopted by many governments.

In an area of biotechnology such as this, it is quite appropriate for ethical discussion of this kind to take place. But I would note something about the 'navigational' strategies that are often used. Commentators need to suppose that the effects promised have already been or can be delivered, otherwise they have no starting point for ethical and political discussion. Cognitive enhancement is an area defined by a lot of hope and promise but little firm evidence. Since it has such strong implications for individual and state competitiveness it certainly merits discussion beyond the 'effectiveness' agenda, but as effectiveness itself cannot easily be established, it needs to be posited. Thus, discussion of mental capital through cognitive enhancement is easily drawn into speculation.

In this chapter I will develop a different approach. Rather than suppose that cognitive enhancing substances or procedures that have reliable and predictable effects now exist or are on their way soon, I pay attention to a present-day attempt practically and in context to make the effect of 'cognitive enhancement' real. Rather than judge cognitive enhancement in the light of ethical principles that have achieved hegemony in political discourse, I consider the uses that individuals and organisations can put the notion of 'cognitive enhancement' to in addressing the practical problems that they encounter. I do not suppose that it is impossible for substances or practices that are intended to enhance cognition and learning to have effects. Rather, I consider the effects that their deployment can have beyond questions of the performance of individuals, focusing instead on how the goal of enhancement affects relations between individuals, families and public and private organisations.

My overall view is that rather than presenting us with reliable and predictable effects, the deployment of cognitive enhancers can define what 'effectiveness' is taken to mean in a given setting by mobilising interests around specific substances or practices. My intention is to shift the critical agenda around cognitive enhancement and, thus, mental capital, away from questions of whether and how the effect of improved individual performance can be achieved and towards consideration of what kinds of 'effectiveness' might be desirable. To do this I will consider an example of an 'enhancement' practice, an attempt to trial 'fish oil' as a way to improve exam results. My arguments will also comprise an illustration of the use of three multiplicities Life, Voice and Resource to shape a distinctive way of navigating this instance of contemporary childhood bio-politics and to open a form of bio-social imagination – mental capital – to critical enquiry.

Mental capital

The UK is a small country in a rapidly changing world. Major challenges such as globalization, the ageing population, the changing nature of work, and changing societal structures are already having profound influences on society and on our place internationally. So, if we are to prosper and flourish in this evolving environment, then it is vital that we make the most of all our resources – and this is as true for our mental resources as material resources.

(Government Office for Science 2008: 4)

In 2008 the UK Government Office for Science published a report called 'Mental Capital and Wellbeing: Making the Most of Ourselves in the 21st Century'. Though the administration has changed since then, the report offers some interesting insights into one state's view of its population in relation to possible futures. Introducing the report, Professor John Beddington, the then chief government scientific officer, has it that the UK population has 'mental resources' that need to be carefully deployed in the national interest. At one level, this is a standard restatement of the basic bio-political governmental approach Foucault (2007) identified. The quotation above, for example, is a clear statement of intent to consider the UK populations' mental capacities as a resource to be fostered and developed so that this 'small country' has an acceptable future place in an international economic order.

The challenges that Beddington refers to are quite specific to contemporary conditions, however. Some are demographic in nature. It is expected that the proportion of older citizens in the UK population will continue to increase in coming decades, as in many other minority world countries, due to a combination of increased longevity and decreasing rates of childbirth. If they are to support others in retirement, the working young adults of the future will need to be highly productive. Other challenges stem from the globalisation of economic activity. It may once have been the case that former colonial and imperial powers, like the United Kingdom, could take their centrality to the global economy and therefore their population's access to work, wealth and well-being for granted. It is clear to Beddington that this is no longer the case. Whether as individuals seeking employment or other contracts with international or overseas businesses, or as employees of private and public concerns, UK citizens will have to compete with others from around the world.

The report identifies a need to increase the population's 'disposition to learn' by intervening in the early years of their lives to give them the ability to adapt, later in life, to changing markets (Government Office for Science 2008: 11). Such figures as 'disposition to learn', 'emotional resilience' and 'mental wellbeing' are then agglomerated within the concept of 'mental capital' which,

> …encompasses a person's cognitive and emotional resources. It includes their cognitive ability, how flexible and efficient they are at learning, and their 'emotional intelligence', such as their social skills and resilience in the face of stress. It therefore conditions how well an individual is able to contribute effectively to society, and also to experience a high personal quality of life.
>
> (ibid: 10)

So with a few deft moves, the report establishes a clear frame of purposes and challenges around the mental powers of the population and creates the figure of 'mental capital', which is at once considered an organising concept of government activity and a new target of intervention. Just like financial capital, mental capital is a resource that needs considered treatment. The report then takes a further step. Based on its survey of existing life science evidence, it asserts that developments in science and technology will present new opportunities for improving the development of mental capital across the life course from early childhood to old age. New scientific understandings of development, learning and learning disabilities are available and the report calls for teachers and front-line childcare professionals to be given scientifically accredited training in these areas. Particular emphasis is placed on intervening in the child's life processes as early as possible. Pre-natal maternal stress, diet, smoking and alcohol use, for example, are identified as high priorities for intervention. The identification and treatment of learning disabilities and the provision of coaching in parenting skills for those parents who have not themselves had skilful parents are also recommended. Since the report addressed 'mental capital' predominantly from the perspective of the life sciences, some of its recommendations escape the normal silos of UK government departments. Alongside recommendations that would affect education, for example, the report also has comments to make about housing:

> Ensuring good housing quality is likely to be important – poor housing is a key factor associated with children's mental development, although it is not known whether the association is causal.
>
> (ibid: 15)

The report is confident that science and technology will be effective in reaching clear educational and developmental goals consistent with the fostering of mental capital, but, mindful of the hegemonic value of 'equality of opportunity', highlights the need to challenge inequalities that may arise in access to the benefits of these developments.

Minds as resources

The report clearly frames UK children's mental abilities as resource. Nevertheless, this surface clarity conceals some ethical and biopolitical complexities. After all, in some contexts and from some points of view, seeing people as resources rather than as 'ends in themselves' is dangerously reductive. It may, for example, appear to give license to the powerful to limit individuals' opportunities for choice. As suggested in Chapter 2, even when this is presented as being for one's own good, it can be reasonable to oppose it. Further, when this 'resource' view is expressed through economic language as in the term 'mental capital', it may raise concerns about the buying and selling of human capabilities and felt identities. In such circumstances, it is possible, if not necessary, to see concerns of resource and of voice as being opposed to one another. Here, the ethical concern would be that the more one appears as a resource, the less one has opportunities to give voice to one's views and preferences.

As I have suggested, the report certainly does see the present and future UK population in a context of economic competition. It also presents good reasons for doing so. On the other hand, it presents little that would directly meet this ethical concern. Beddington does offer the following, however;

> The idea of 'capital' naturally sparks association with ideas of financial capital and it is both challenging and natural to think of the mind in this way.

(ibid: 15)

This appears at once to acknowledge the ethical and political resource/voice tension but also to set it firmly aside. It is striking that this issue is addressed by the assertion of the 'natural' quality of the 'mental capital' frame. Given that no evidence is presented to support this assertion, we might wonder where the confidence with which it is made comes from. One answer is that the multiplicity of life is drawn on to provide contextual support for something that would otherwise be quite contentious. The report is not just concerned with economic prosperity but also with 'quality of life'. It frequently refers to the 'flourishing' of the population. It is concerned to challenge disadvantage and inequality by using life scientific insights to enable children to reach their future potential. Thus, even as an economic assessment of the worth of individuals is deployed, the report presents us with the opportunity to see this in the broader context of a bid to enable a more fundamental liberation. Human flourishing is the release of 'potential' understood as innate and natural. Concerns for quality of life thus enable the report to make the mental capital frame 'natural' by presenting it as consistent with a supposedly incontestable desire to enable the natural flourishing of human capacities. This is quite an effective strategy, at least on the page. Making the most of ourselves in the neoliberal near future envisaged will involve high productivity and the mental strength and agility to compete. If this conjures up images of unstable, highly demanding working conditions that crowd out other aspects of citizenship and family life, we can nevertheless accommodate ourselves to this if we are prepared to accept that we are on a path to becoming the best that we can be.

In sum, where conflict or opposition threaten to arise at a local meeting point of multiplicities of resource and voice, the attempt is made to neutralise this by moving accounts to the third multiplicity of life and the notion of 'flourishing'. Whatever effects of consistency, clarity and reason this produces depend on the ambiguity of fact and value that is shot through the multiplicity of life. This provides the report with the opportunity to use life scientific findings to substantiate and to naturalise the otherwise vague and value-laden concept of 'flourishing'.

Ambiguities of enhancement

If a universal species-normal level of human cognitive powers or learning capacity could be established, it would be relatively easy to define cognitive enhancement. Any measureable increase above this 'normal' level would count. However, human cognition and learning do not spring from the life processes contained within individual central nervous systems alone (Vygotsky 1986). Historically, these capabilities have also been variously conditioned, repurposed, diversified, concentrated and restricted within political, social and cultural processes. Thus, individual humans certainly can be seen as having levels of individual intelligence related to the life processes of their individual central nervous systems, but they can equally well be seen as having socially produced and culturally conditioned 'extelligence' (Stewart and Cohen 1997). Cognitive powers have also been amplified and diversified by technological developments, like writing. On this view, it is far from obvious that there is a universal species-normal level of cognitive powers and, further, thinking of these powers as the stable and measurable property of individuals begins to seem questionable.

These observations do not lead me to the view that there is no such thing as 'cognitive enhancement' or that substances cannot affect cognitive function, but to the view that the concept of 'enhancement' is fundamentally ambiguous. One can take steps to establish a baseline for an individual or for a population, but these are local to the context of the studies they are deployed within. On this view, the clinical and non-clinical trials that have been deployed to test the enhancing powers of drugs cannot so much establish a generalisable truth about the powers of a substance, as build highly unusual settings that temporarily set the ambiguity of 'enhancement' to one side. This does not stop cognitive enhancement from organising biosocial imaginations about individual life processes and fitness for social and economic competition. Indeed, as I will shortly illustrate, its very vagueness can be a spur and an excitement. My core concern in what follows is with the restriction of bio-social imaginations that the mental capital frame can perform. The more successful it is in defining human potential on the model of capital, the fewer opportunities there are to grow other ways of thinking about human capabilities in relation to the world, other ways of being 'effective'. I'm concerned here by the idea that from some powerful points of view it is already clear what human potential for growth and change in conduct and in relation to the world is. I am further concerned that, in the context of cognitive enhancement debates, stronger individual powers of concentration, more accurate memory and faster information processing are treated as an acceptable proxy measure of the fulfilment of such potential.

Fish oil and the ambiguity of enhancement

Fish oil dietary supplements, such as cod-liver oil, have long been credited with health-giving properties. Fish also enjoys a reputation as 'brain food'. In recent years, a version of the mismatch thesis described in Chapter 3 has been advanced, linking fish oil dietary supplements with cognitive performance. This account builds on the idea that significant periods of human evolution took place near supplies of seafood, such that our bodies and minds now have certain dietary requirements. There is a class of bio-chemicals called essential fatty acids (EFA) which are needed in a range of human life processes but which, like other mammals, humans are unable to synthesise within their own bodies. We are thus dependent on the food we ingest to supply both an omega-6 EFA called linoleic acid and an omega-3 EFA called alpha-linoleic acid. Omega-6 is available in wheat and vegetable oils like palm, sunflower and rapeseed. Omega-3 is available from fish along with other sources. 'Western' diets that are built on wheat and vegetable oil and that carry relatively little fish contain omega-6 and omega-3 in a 15:1 ratio. Estimates are that humans evolved in environments that provided a 1:1 ratio of these kinds of EFA (Simopoulos 2002). This difference of ratios is understood to be one factor influencing rates of arthritis, inflammatory and autoimmune disease and cardiovascular disease. Given this background, there is considerable research interest in exploring the effects of changing this ratio in favour of omega-3 on the cognitive performance both of clinical (Sydenham et al. 2012) and of otherwise healthy populations (Office of Science and Technology 2005).

In this chain of associations, where some links are proverbial and some bio-scientific, we have a good example of the ambiguity of 'cognitive enhancement'. If we suppose that the mismatch account is correct, then the individual cognitive performance of many present-day humans will be in some sense lower than that of their ancestors who enjoyed a diet rich in seafood. In that case, a large number of present-day adults and children are, considered on an evolutionary time scale, in a state of relative cognitive deficit. Given all this, changing their EFA ratios in favour of omega-3 cannot be properly be said to 'enhance' them, but only to return them to a species normal state. However, if we imagine a situation in which the local individual normal performance is set at the levels we would expect to result from an omega-3 poor diet, then giving some rather than others extra omega-3 will, presuming a real enhancement effect, create a relatively enhanced group when compared to their peers. So, even if the mismatch account were proven, there is scope for fish oils both to be considered 'remedial' and to be considered 'enhancing'.

As things stand, there is no scientific consensus that fish oils can change cognitive performance in clinical or healthy populations. More importantly, from my point of view, it is clear that which context we are to take as 'normal' such that we may detect 'enhancement' is a matter of decision not a matter of fact. As the following example suggests, this quality of ambiguity enables the figure of cognitive enhancement to contribute to the cementing of a definition of 'effectiveness' in human cognition and in education as the passing of exams by individual pupils. As it does so, this element of the multiplicity of life helps to reconfigure understandings of children as resource for themselves, their families and education authorities and offers them new, but tightly constrained, opportunities for voice.

The Durham fish oil trial

Unlike many other substances that are currently understood to have potential as cognitive enhancers, fish oil supplements are relatively inexpensive, widely available and can be taken without medical prescription (Office of Science and Technology 2005). In 2007, Durham County Council in the United Kingdom took advantage of this combination of the promise of enhancement and the opportunity fish oil supplements presented. Using products supplied at no cost by the company Equazen, it attempted to test the ability of fish oil supplements to increase school pupils' performance in a set of public examinations taken by 16-year-olds and known as the General Certificate of Secondary Education (GCSE). In the United Kingdom the proportion of pupils gaining C grades or above in five different GCSE subjects is used as a key means for central government to assess the quality of education being delivered at individual schools and in local authority regions. This meant that Durham was constrained to see effectiveness in terms of GCSE performance and, further, was motivated to enhance its pupils' performance.

The Durham trial was an attempt to discover evidence for a causal link between the presence of a fish oil supplement in pupils' diets and their GCSE exam results. At its inception the trial had some 3,000 pupils of 15 and 16 years old taking supplements at home and at school in the run-up to their exams. At exam time 800 pupils were still taking them. In search of evidence to inform the question of the effectiveness of fish oils, Durham Council's Children and Young People's Services division compared the GCSE results of those who had remained compliant throughout the trial with those of children who had not taken the fish oil supplement. The performance of the two groups did differ. Those who were compliant throughout the trial had better results than those who had not taken the supplement. But did this difference constitute evidence of a causal enhancing link?

The trial received a good deal of press coverage not all of which was positive. The science journalist Ben Goldacre (2009) used it as an example of 'bad science'. On his view, the Durham trial was, from a scientific point of view, a failure in the sense that its faulty design meant that whatever results it yielded could not be considered meaningful. Differences in performance between two groups were observed. However, there was no reason to attribute these differences to the activity of fish oil within the pupils' nervous systems. The compliant group who had taken the oil throughout the trial were self-selected and gave signs of being highly invested in educational achievement. There was no way to tell, within the Durham design, whether the fish oil or the participants' investment in education lay behind the differences in performance. The trial made no attempt to control for placebo effects. Finally, no information was sought or fed into the trial design and interpretation of result about the diets and supplement use of children who did not take the Equazen product. It is quite possible, therefore, that children who scored relatively poorly in their GCSEs were already taking fish oil supplements independent of the trial or had diets rich in EFAs. So the question of effectiveness was thoroughly fudged in the Durham trial. That, however, does not mean to say that the trial was without effect.

Responding to some of the negative press coverage of the trial, the Head of Achievement for Durham Children and Young People's Services issued a press release (Durham County Council 2008). He acknowledged the faults in the trial's design and accepted that they mean that no positive inferences about the effectiveness of fish oil could be made. He also pointed out, however, that if no difference in GCSE performance had been detected between the two groups, Durham County Council would have dropped the issue entirely. This led him to the following;

> ... taking all this into account, it is our view that this study has produced some interesting and possibly exciting issues for further investigation that could be the basis for future scientific trials ....
>
> (Durham County Council 2008)

So an imaginary negative result and a meaningless positive result combined gave him reason to remain hopeful about fish oil.

Whether or not the Durham trial was 'bad science' and whether fish oils are effective as enhancers are important questions. But they are not my concern here. Instead I'd observe that in these circumstances, the idea of cognitive enhancement sat within a frame of mental capital and cemented a definition of learning effectiveness as success in examinations. Along the way it helped to organise a coalition of interests that shared a view of pupils as resource while holding different understandings of pupils' value. For Equazen, the Durham pupils were a proxy for future consumers of their products. Whether or not the trial was successful, they stood to gain publicity for their products. For Durham County Council, pupils were a resource that needed to be fostered if it was to meet its performance targets and thereby enhance its own reputation and powers. For families and pupils, fish oil presented an opportunity to take deliberate steps to intervene in their own future success, as currently defined by educational institutions, treating the functioning of their nervous system as a resource to be fostered. All this could take place in the absence of general evidence of fish oil effectiveness and in the presence of a poorly designed study.

Voice and pharmaceutical enhancement

The use of fish oil in the Durham trial may well have appeared low-risk to participants. EFAs, after all, are standard foodstuffs and Equazen products are part of a relatively unregulated and thriving market of non-prescription boosters and supplements addressing such phenomena as 'immunity' and 'mood' (Coppens et al. 2006). As I indicated above, however, there is growing interest in the use of more tightly regulated pharmaceutical products. Ritalin, for example, offers young people a clear pathway to voice and agency, promising them greater traction over their own academic performance and their futures in an economically competitive world. Developed and prescribed for Attention Deficit Hyperactivity Disorder, Ritalin is now widely used by students in attempts to boost their powers of concentration and intellectual performance under pressure (Harris 2007).

As I have argued, the range of opportunities for children and young people to articulate their views and preferences concerning 'enhancers' are at present rather narrow. As long as one sits within the mental capital frame, once can choose to use a given enhancer, to find other ways to boost one's brainpower or trust to chance and innate ability. There is scope, however, for more complex articulacy about enhancement to develop even within this frame. There are opportunities, for example, for young users of enhancers to share subjective information about the nature of effects and side effects, how they decide upon dose levels, and what metrics and information they use to decide on the success or failure of a given enhancement attempt. Voicing effects like these may be shaped by supply chains, by peers and by Internet communities (Coveney et al. 2011). As I noted above, scientific trials of enhancers do not so much resolve ambiguities about effectiveness as set them temporarily to one side. Looking ahead, it will be interesting to see whether and how children and young people's discussions of enhancement draw on scientific authority and whether and how they manage the ambiguities that enhancement raises.

Conclusion

Enhancement practices can be understood as attempts to realise the desire to create reliable and repeatable bio-social events so as to draw the life processes that comprise human minds ever more closely into resource status. This desire can be held by children, caregivers, communities, governmental agencies and business independently and shared between them. Whether or not fish oils or Ritalin are effective in conditioning cognitive function, the idea of enhancement certainly is effective in cementing a view of what an effective child is. A view of substance effectiveness as the ability to boost exam performance is consistent with the mental capital frame. Defining the effectiveness of children and young people in terms of passing public examinations is also consistent with this frame.

Two questions then present themselves. First, is the mental capital frame consistent with the lives and experiences of all children, families and communities? Not all individuals, whether adults or children, are centrally concerned with becoming economically productive. The overlap between governmental concerns and those of the population is far from complete. I would read the concerns for equality of opportunity that cropped up fairly often in the above discussion to be a partial recognition of and an attempt to respond to this. Second, however, is the more complex question of whether the mental capital frame is consistent with the demands that life is going to place on children and young people as they age. There is some tension here between different uses of the mental capital frame. When mental capital is presented through issues of cognitive enhancement it seems very closely tied to a narrow understanding of effectiveness in learning and what it is to be an effective young person. Arguably, there is an affinity between the promise of cognitive enhancement and the kinds of activities that are involved in passing public examinations where good recall, focus and concentration are rewarded. In the report, however, the frame presents as a general concern to foster cognitive powers, whatever form they take, in the name of future flexibility and adaptiveness in the workforce.

Sadly, it is becoming very clear to many European school leavers and university graduates that while success in public examinations is valuable, it is no guarantee of employment. The current European recession is just the kind of economic shock that the authors of the report wanted to prepare young people for. It seems that their sense of impending economic instability was just right, but that, unfortunately, they were a little too late in giving their advice.

What other views might there be of 'effectiveness' in learning and in the use of human mental abilities? The currently dominant bio-social imagination that finds cognitive enhancement so interesting and a view of mind as capital so 'natural' has a markedly anthropocentric view of resource, one in which humans beings are the principal resources in a game of international competition to grow economies. The mental capital frame is haunted by the fear that human intelligence might be in short supply relative to need. A broader view would take other resources and forms of resource scarcity into account. In later chapters as I turn to the topic of climate change I will argue that we have choices about how we understand human cognitive powers and agency. If we narrate human history anthropocentrically as an increasing separation from other life processes that can reliably be arranged to our convenience, we end up dreaming of changing our own biochemistry to make ourselves 'better'. If, on the other hand, we narrate human history in the light of our dependence on scarce resources that are provided by a wider matrix of life processes a quite different set of ideas about effective education and human intelligence and agency becomes resonant. I will discuss this alternative through the concept of 're-framing', an active and reflective form of bio-social imagination.

5
Childhood, Vaccination and Philanthrocapitalism

Historically, children have been the focus of many philanthropic and charitable projects. These have been undertaken for diverse motives and have had a range of outcomes. In some cases, NGOs have emerged from early philanthropic roots to continue to work for children's well-being. The Barnado's organisation (www.barnados.org.uk) is a good example of this. In 1868 a Dr. Barnado established the 'East End Juvenile Mission' as part of a wider campaign for the care of deprived and destitute children in UK cities. The mission clothed, fed, deloused, medically treated and housed children while his campaign made extensive use of 'before and after' photographs of them to raise awareness and funds. Barnado was motivated in these efforts to promote children's well-being by his Christian faith, which continues to inform the work of the organisation that still bears his name.

Consider, however, the case of a Barnado's boy, Peter Fortune. Born in 1943, at age 11 he was selected for a migration programme organised by Barnado's in cooperation with the British state. Like thousands of other children across the mid-twentieth century, some orphans and some not, who were transported from the United Kingdom to commonwealth countries, Peter was sent to Australia to live, learn and grow up. Barnado's had a long history of placing children in Australia, New Zealand and Canada and had strengthened its relationship with the British state in preserving and managing the resource of the young during the Second World War when it helped organise children's evacuation from major UK cities. From the point of view of those making decisions to transport children like Peter, the

United Kingdom had a surplus of orphaned and destitute children, while commonwealth countries needed more young workers. Further, in their view, a new country would mean a better future for the children than either their parents, communities or British society as a whole could provide.

Like many of his peers, Peter was housed in a 'farm school' where he was set to agricultural labour. Sexual abuse of the children by staff is known to have occurred there and Peter is thought to have been abused. He later went to live with a family and eventually married the daughter. In 2003, however, he was convicted of sexual offences against 12 women. In 2006, having served a prison sentence, steps were taken to deport him to the United Kingdom, leaving his wife and everyone he knew behind in Australia. Even though he was not an entirely sympathetic character, you might think that having been transported as a child and subsequently lived in Australia for many years, Mr. Fortune would be treated as an Australian citizen. But he, like many others, was deemed to have retained his UK citizenship, setting limits to Australia's responsibility for him. There is a growing literature describing similar experiences (Gill 1998). Mr. Fortune's story makes it clear that positive, philanthropic intentions do not ensure positive outcomes.

I do not recount this story to single out Barnado's or any government of the past or present for criticism. It does show, however, that children are often used as resources in advancing the goals of states and of other organisations and, further, that states, philanthropic and charitable organisations do not always act responsibly towards the children they use or with due care towards the relationships that they create. I can think of no reason to suppose that these matters have come to a full stop, or that the present and future are likely to be radically different in this respect from the past. After all, children are a primary source of value within human societies, their birth providing the biological resources that all societies need to continue, to grow and to adapt. Deploying children as a resource is not necessarily illegitimate, but cases like Mr. Fortune's should orient us to the question of how to respond to the issues of responsibility that it poses: If positive intentions have soured in the past, what caution should we apply today and where do we need to watch our step?

Mr. Fortune's story shows that outcomes for children who are the object of philanthropy are shaped, in part, by the nature of the relationships that are created within philanthropic and charitable projects between wealthy or charismatic individuals, the organisations they found or work within and states. Peter's transportation and deployment seemed to solve demographic problems of state strategy – surplus and shortage of young people for the United Kingdom and Australia. But another problem was created – Peter's vulnerability to abuse. Arguably, the philanthropic history and traditions of Barnado's, by casting his transportation as fundamentally in his interest, contributed to his vulnerability to abuse by those tasked with his care. Who would suspect or challenge individuals who were working in partnership with a charity that had been devoted to saving children's lives, preserving their well-being and helping them reach their potential for almost a century? Likewise, the specific relations of personal, democratic, media and legal account making and accountability that underlie the metaphor of 'voice' matter a great deal here. The fact that there can be tensions between resource and voice, and that these get played out over the lives of individual children is easily seen in terms of the many occasions on which Peter could have been consulted about his preferences and concerns, but was not.

Since the widespread institutionalised abuse of children came to light in the late twentieth century, charitable and state guardians of children's welfare have attracted increasingly close inspection and regulation. Hearing children's voices is a big part of this. As I hope to show in this chapter, however, good intentions to save children's lives preserve their well-being and to help them reach their potential can still be advanced as a guarantee of positive outcomes and as an assurance against harm. Further, today we need to consider a wider play of 'voice' alongside that involving individual children like Peter, and more complex articulations of account making and accountability that are accompanied by sophisticated structures of silencing. Later in the chapter I will advance this view by examining material drawn from a televised debate between the philanthropist Bill Gates and two development workers. In my view we need to pay particular attention to those circumstances in which philanthropy relates to children's lives, well-being and potential as if they were 'beyond politics'. I am probably not alone in having positive feelings about any actions that are taken to save children's lives. However, as I will argue, there is a cost to presenting philanthropic work with children as if it were a simple matter of life or death.

Wealth and global elites

By some estimates, the global population has become much wealthier in the past few decades. One figure reported is that total global wealth increased from $195 trillion in year 2000 to $231 trillion in 2010 (Credit-Suisse 2011). It is certainly the case that, thanks to a new wave of industrialisation and urbanisation across the developing world led by increased globalisation of trade, more individuals than ever before now participate in the money economy and do so more intensively, so have greater measureable assets. Think of a Chinese mother who has migrated from a region of subsistence agriculture to work in a factory in the Shenzhen region leaving her child behind with grandparents. She now has a wage, where once money was much harder to access (Ngai 2005).

Not all definitions of a good life centre on such participation or on the quantity of money that is being generated, however (Jackson 2011). Other forms of prosperity exist, such as access to fertile land and water and basic freedoms from the political violence and economic oppression that degrade people's ability to support themselves through their own labour. But individuals' and communities' access to the basic means of self-support and the monetisation of economies are often in tension. This is because, whether it takes the form of credit advanced by a bank or personal savings, financial liquidity ultimately derives from the basic economic activity of converting natural resources into tradeable commodities, such as wheat, palm oil, soya beans and cotton and products like clothing and food. This may be a wealthier world, but it is also one in which inequalities in the distribution of wealth and other forms of prosperity have arisen. The means by which natural resources are converted into money through labour and trade play a fundamental role in this.

Consider the differences between the situation of a dozen independent subsistence farmers and 11 farm labourers who are working for one farm owner. The latter operation has access to certain efficiencies. First, a large farm owned by one individual is, to a bank, a more attractive proposition as collateral for a loan than many small plots. It is easier to negotiate a contract with one individual than with many and, in the event of foreclosure, the transaction costs of selling the land to recoup losses are proportionally smaller as the size of the property increases. Second, borrowed money can be invested in machinery to increase the farm's efficiency.

One benefit that employees have that independent subsistence farmers do not enjoy is access to money in the form of a wage. However, they also have to negotiate their rate of pay with the landowner who is in a position always to pay them less than the value that their labour actually generates. They also then depend on the landowner and his or her investment decisions for their well-being. The same relations hold as we ascend the 'value chain' from farmers to major food distributors like Nestlé and Kraft or from miners and factory workers to information and communications technology companies like Microsoft, Apple and Google. Without a well-developed value chain, it would be very difficult to bring many useful and innovative products to market. At the same time, the better a value chain is at delivering products, the more likely it is also to concentrate that money on relatively few individuals, whether they be shareholders in a range of corporations, or the figurehead of just one.

These are some of the processes that have led to the contemporary level of global wealth and the patterns of its distribution among nations and individuals. As I have suggested, there is a tendency for processes of wealth generation to lead to concentration of wealth in relatively few hands. This has been especially pronounced over the past few decades because relatively wealthy sections of the population have been able to influence national and international political decisions about the regulation of trade, investment and banking in their favour. Thus, the world has certainly become 'richer' in the sense both that more people and more natural resources have become involved in value chains and in the sense that more money has become available for investment and expenditure. But, at the same time, a large proportion of that global wealth has become so concentrated that many commentators now refer to a global elite class (Freeland 2011).

Philanthrocapitalism

Bill Gates, the co-founder and chairman of the software company Microsoft, is a prominent contemporary philanthropist. His personal wealth varies with fluctuations in the share value of the company Microsoft, but it has been estimated at around $66 billion dollars (Forbes 2012), making him one of the world's richest individuals. Microsoft products have provided the data-processing infrastructure that has enabled recent globalisation of trade and the fresh wave of industrialisation in the developing world.

Bill Gates has been involved in philanthropic work at least since 1994, and formed the Bill and Melinda Gates Foundation (henceforth 'the Foundation') in 2000 by merging the William H. Gates Foundation and Gates Learning Foundation. More recently the investor Warren Buffett, once ranked as the world's richest individual, has made his own contribution to the Foundation. These foundations have, since their inception, distributed over $25 billion dollars to projects across the world. The annual budget that the new Foundation makes available to its Global Health Programme is around $800 million dollars. This figure nearly matches the budget of the United Nations World Health Organization that derives its funds from the whole world's national governments. This makes the Foundation a major actor in health policy and provision around the world and places its figurehead and chief benefactors on a par with, or even some way ahead of, the elected leaders of most countries as an influence in this area.

Just as Dr. Barnado's philanthropy was shaped by his religious faith and the visibility of child poverty in late-nineteenth-century UK cities, so the Foundation's activities have been shaped by their cultural and economic context. For Barnado, the correct response to human suffering was to emulate Christ through charitable action and to encourage others so to do. Beyond children's material needs and physical survival, he took aim at their spiritual condition. As he wrote in the 1860s,

> ... the chief aim of all associated with me (irrespective of Churches or denominations) is to bring these children up ... in the fear of the Lord, and to draw them in faith and love to the feet of our Saviour Christ.
>
> (Barnados 2013)

Compare that with the following extracts from the Foundation's guiding principles.

> Guiding principle #2 Science and technology have great potential to improve lives around the world.... #3 We are funders and shapers – we rely on others to act and implement.... #4 Our focus

is clear – and limited – and prioritizes some of the most neglected issues. . . . #5 We identify a specific point of intervention and apply our efforts against a theory of change. . . . #6 We take risks, make big bets, and move with urgency. We are in it for the long haul.

(www.gatesfoundation.org)

Three features in particular are key to the Foundation's approach; the use of scientific and technological means to improve life; the clarity with which it defines its own role and selects and prioritises the issues that it will address; and, the commitment to monitoring the success of its efforts. The Foundation's guiding principles, then, embody values that concern the clarity and transparency of decision-making processes. The Foundation construes philanthropy as a matter of taking risks and making bets, hoping to make big advances in areas that other organisations have not been able to tackle. Transparent decision-making and monitoring of outcomes provide the confidence to take such big risks.

In these respects the Foundation's approach appears to be modelled on the investment practice of 'venture capitalism'. Here, wealthy individuals seek out opportunities to invest in innovative young businesses that have the potential to be highly profitable but which cannot yet access capital through a bank loan or other means precisely because they are young, innovative and, thus, a risky investment. The venture capitalist provides the resources rapidly to develop the business to fulfil its potential and is rewarded for investment through part-ownership of the business. Venture capitalists accept that some investments will fail. Individual investments are often structured so as to minimise the costs of failure to the investor. The investor, holding the money, has a powerful position in any such negotiations. But across a series of investments, the venture capitalist carefully monitors their own decision-making and the roles they take up so as to keep the success/failure ratio in their favour.

The term 'philanthrocapitalism' has been coined to describe this application of the habits and values of venture capitalists to philanthropic work (Bishop and Green 2010). State and charity funded development and emergency relief work are never without their critics (Rapley 2004). Given this, philanthrocapitalism could be viewed simply as the application of sound business discipline to make good the limitations of state and charity funded development and emergency relief work. But just as state aid efforts reflect their origins in colonial history (ibid), philanthrocapitalism is a reflection of the conditions under which today's global elite emerged. As I will argue shortly, this continues to affect the relationships created within philanthrocapitalism, raising questions of accountability and responsibility.

Preventable disease and the Global Alliance for Vaccines and Immunisation (GAVI)

Children worldwide are vulnerable to a wide range of infectious diseases. The Rotavirus pathogen is the most common cause of diarrhoea, infection by Pneumonoccocal bacteria can result in pneumonia, blood infection and bacterial meningitis, and malaria is an infection of the blood by Plasmodium parasites that causes recurrent fevers. Whether one of these infections kills an individual child or not depends on a range of factors, but the child's general levels of health and nutrition and their ability to access medical care are critical. Across the majority world these protections are not always available. Thus, Rotavirus kills about 450,000 children through dehydration per year and affects countless others, while malaria causes an estimated million annual deaths and 250 million cases of severe periodic fever. As noted in the United Nations' Millenium Development Goals, much of this illness and death is preventable.

Given that it can be very difficult to remove pathogens from children's environments, the preparation of children's immune systems to work effectively against them through vaccination has been identified as the most effective available response. Rotavirus and Pneumonoccocus vaccines have been available for many years. However, many developing countries have been unable to afford those treatments at the prices charged by the pharmaceutical companies that manufacture them. These prices are not set arbitrarily. The laboratory and field research that go into creating a safe and effective vaccine carry significant costs. There is no point in the chain of development, testing and deployment of vaccines at which costs are zero. Unit prices for vaccines are also boosted by developers' uncertainty about whether governments in developing countries are able or willing to make a long-term commitment to purchasing them. Perceptions of countries' political instability and levels of financial distress thus affect their children's access to immunisation against preventable disease because developers often deal with risk by raising prices.

The Global Alliance for Vaccines and Immunisation (www. gavialliance.org) is a group of organisations committed to overcoming the mismatch between poor children and expensive vaccines. Its governing board comprises representatives from WHO, UNICEF, the World Bank, governments of aid giving and receiving countries, the vaccine development and production industry and the Bill and Melinda Gates Foundation. About three quarters of its funding comes from governments and a quarter from private donors like the Foundation and businesses. GAVI has a very clear picture of the challenge it is taking on: There are preventable infectious diseases that are killing children. Pharmaceutical companies have the biotechnological capacity to provide vaccinations against them, but development and production costs are high and their risk management strategies also raise prices. Many governments that would wish to vaccinate their children cannot afford to pay. GAVI thus construes the preventable sickness and death of majority world children as a case of market failure. Under current circumstances, the value chains that should produce useful and innovative products for customers along with running costs and profits for industry are not functioning as we might wish. The results are that potential customers are not being served and pharmaceutical companies' potential markets are constrained.

One way for a philanthropist to intervene in a situation like this would be simply to provide governments with the money to buy vaccines for their children at the price currently demanded by pharmaceutical companies. As long as the money lasted this would provide a flow of vaccines to the point of greatest need, provided, of course, that these large transfers could be secured against corrupt diversion. However, even if the capital behind such a fund were never spent, the effect it could have would be limited to the level of interest or other investment return accruing on that capital. All the while, the basic market failure would be left intact, presenting future generations with a continuing legacy of mismatch between poor children and expensive vaccines. GAVI does distribute funds in this way, positioning itself as a buyer of last resort when necessary, but does so in the context of a broader approach to managing risk perceptions

and price. Rather than simply absorb the price for vaccines that the
market currently generates, effectively making donations to pharmaceutical companies, GAVI attempts to build capacity for the use,
development and administration of vaccines and, crucially, to shape
the market by altering the relationships of cost, risk and trust that
inform decision-making by pharmaceutical companies, governments
and others. It expresses its strategy in terms of four goals, each of
which has clear metrics of success or failure attached:

> The vaccine goal
> To accelerate the uptake and use of underused and new vaccines
> The health systems goal
> Strengthen capacity of health systems to deliver immunization
> The financing goal
> Increase predictability and sustainability of financing for
>   immunization
> The market-shaping goal
> Shape vaccine markets to provide affordable and appropriate
>   vaccines

> http://www.gavialliance.org/about/strategy/

Over the five years from 2011, GAVI aims at 100 new vaccine intro-
ductions, mostly against Pneumonoccocus and Rotavirus. It will
continue to build capacity in the health systems that will actually
immunise children by increasing civil society involvement in building and running those systems and making incentive payments to
governments that reach immunisation targets. The financing goal
means securing long-term and predictable demand for vaccines. This
will involve making sure that pharmaceutical companies are confident that when they invest in the skills and plant to make, say
Rotavirus vaccines, their products will indeed be purchased further
down the line. GAVI's public and private donors have created a large
fund that offers just this confidence.

The market-shaping goal is, perhaps, GAVI's most innovative contribution. GAVI is an alliance of vaccine producers and consumers.
Normally the communication between producers and consumers
takes place through levels of price and purchase. As we have already
seen, communication through these routes has so far been unable

to resolve the contradiction between high levels of biotechnological knowledge and high rates of preventable death and disease among poor children. GAVI is in the unique position of being able to pool knowledge about existing and future need for vaccines and pass that knowledge to vaccine producers. This enables pharmaceutical corporations to see the diseases of poor children in a new light – not as a risky and unprofitable market in which high unit costs are the best way to insulate themselves against risk, but as a relatively low-risk market in which lower unit prices are sustainable. Greater confidence in future demand leads to lowered unit costs. There is evidence that GAVI has already been effective in shaping markets in this way. In 2010 UNICEF persuaded pharmaceutical corporations to release potentially sensitive records of the prices they have charged over the last decade. These reveal that price per dose of hepatitis B vaccine (DTP-hepB and pentavalent) has fallen from $0.59 to $0.20 over the past decade.

Assessing philanthrocapitalism

Through GAVI and other initiatives, the Foundation has pioneered philanthrocapitalism and, judging by the plentiful evidence they publish, is doing a great deal of good. The record of donation is impressive, as are many of the outcomes. Together they provide a strong indication that the insights of people like Bill Gates and Warren Buffett, who have become hugely wealthy by understanding and shaping markets, provide a valuable new angle on problems of uneven global development. They are reframing the issue of children's health in relation to vaccination. In the past, sovereign states stood alone in their negotiations with pharmaceutical companies. It was difficult to identify and mobilise common interests between the need to preserve children's health and the need to balance the books of commercial organisations. Today, a stubborn mismatch between poor children and expensive vaccines has become newly tractable thanks to a re-imagining of the relationship between commerce and children's health. In this sense, philanthrocapitalism is an example of the productive capability of new bio-social imaginations.

Instead of simply trying to aid and persuade majority world governments to change their behaviour towards their populations, through

GAVI, the Foundation aims at shaping the entire nexus of issues comprised by government stability and expenditure and pharmaceutical corporations' abilities and needs so as to drive price reductions and to make long-term commitments to serving children possible. This is just what one would hope to happen when philanthropy is undertaken by individuals who have prospered through the calculation and the strategic manipulation of price, risk and investment. Gates and Buffett's examples are encouraging other members of the global elite to apply their wealth and expertise to development issues (Bishop and Green 2010). The Clinton Global Initiative, for example, led by former President Bill Clinton is in the same philanthrocapitalist business.

Throughout the chapter so far I have implied a comparison between two examples of philanthropy. Barnado's nineteenth-century efforts took aim at the lives and the souls of poor children. Even though it is notoriously difficult to measure souls and their proximity to salvation (Weber 2010), the good intention this represented provided credible account of activities that otherwise might have raised questions – the separation of children from their families and communities and their transportation from the United Kingdom to other commonwealth countries. Gates' philanthropy is quite different in intention and methods. First, it has exclusively earthly objectives and takes a pride in the quantitative monitoring of its own performance. You do not need to believe in a Christian God to be persuaded of the Foundation's or GAVI's case because they provide you with the metrics and the data by which they wish to be judged. Second, Barnado's activities took place within the confines of nation states and came to be significantly shaped by national interests. He depended on a greater power to act as a lever to increase the scope of his work as this grew from city-based children's homes to international migration. In comparison, Gate's Foundation has resources that rival those of nations and international organisations. This means that it is able to operate independently of the states, businesses and NGOs that it shares the GAVI board with. Recall that only a quarter of GAVI's funds come from private sources. Arguably, this means that the Foundation is using its relatively small resources to lever donor countries towards a philanthrocapitalist approach to development. As well as reshaping markets, the Foundation would seem to have the power to reshape understandings of good developmental and philanthropic practice along with the basic questions of what needs fixing and how to fix it.

I have also drawn on the experience of Peter Fortune as a warning against simply accepting philanthropists' own assessments of their achievements and effects. There is no doubt that Barnado's did a lot of good for children over the course of the twentieth century. But for many years it remained insensitive to some of the suffering it fostered by creating new relationships between isolated children and abusive caretakers. So, in my view there is a need to retain a critical voice even in the light of GAVI and the Foundation's evident success. In this context, being 'critical' certainly does not mean stubbornly opposing philanthrocapitalism. It would, in my view, be absurd simply to decry successful efforts to save children's lives. In my research for this chapter, I have been unable to find any commentators who do so. Likewise, a piecemeal critique comparing GAVI point by point against some imaginary perfect form of global health intervention would simply make the 'best' the enemy of the 'good'.

As I have suggested however, a key feature of philanthrocapitalism is that it operates on the basis of a clear view of the nature and limits of the problem, the limited role of the philanthropist and the proper metrics of success and failure. Much of its strength derives from this focus and much of its credibility rests on the impression of accountability this creates. Over the past decade or so, philanthrocapitalism has been very good at 'framing' the issue of children's health and potential in its own terms. In this sense it is a kind of bio-social imagination that weaves life and health together with the social and economic relations of risk peculiar to pharmaceutical businesses and governments. In my view this strength and credibility calls for a critical perspective for the following reasons;

- There is always more than one way of framing an issue and every frame has its special forms of sensitivity and ignorance
- Successful philanthrocapitalism may be changing what 'health' means in relation to international development and this merits examination
- Others, perhaps less capable and circumspect than Bill Gates, are likely to follow his lead.

Reframing health inequalities

One way to create a critical perspective on a very successful frame is to compare it with an alternative. As I have argued, GAVI frames the health and well-being of children in terms of the successful or unsuccessful operation of markets for vaccines. On this view, under normal conditions, the production cost of vaccines and the ability of their consumers to pay for them end up well matched. Mismatches between cost and ability to pay represent exceptional cases, the failure of markets that can usually be relied upon to deliver benefits both to producer and to consumer. Paul Farmer (2004) is a medic and anthropologist whose practical experience has led him to develop his own frame for understanding relations between poverty and disease. The question that draws his attention is one that does not appear on Foundation or GAVI websites: Why does preventable disease affect the lives of so many throughout the majority world? For him, the global distribution of the burden of disease that sees poor children dying from conditions, like diarrhoea, that wealthier children would easily survive is one of the many forms that political and economic oppression takes. It is not simply unfortunate that some children die of diseases that are preventable or treatable, it is the result of decisions made and actions undertaken by powerful individuals and groups.

I can illustrate Farmer's frame with the case of the Democratic Republic of Congo (DRC). The land of the DRC contains great mineral wealth. There are deposits of the metals gold, tin and tungsten and the compound columbite-tantalite, often referred to as 'coltan'. Each of these has uses in the manufacture of consumer electronics – computers, mobile phones and so on. Between August 1998 and July 2003, the Second Congo War, also known as the Coltan War involved eight nations, and resulted on the deaths of millions. One of the matters fought over was access to mineral deposits. Another was access to adults and children to press into slave labour to mine those deposits. Those in command of military forces were in a position to profit hugely from the pursuit of this war. By forcibly turning land into mines, and farmers and children into miners, they could enter the 'value chain' above their captives, pay them nothing, smuggle the minerals across DRC borders to render them saleable within respectable global commodity markets. Without this, many of the mobile phones and data contracts bought and sold over the past decade could simply not have been produced.

From the perspectives common in the wealthier parts of the world, it often seems that the disease and death of children are best understood as biomedical matters. Within the self-imposed limits of philanthrocapitalist frames, markets are understood to benefit producer and consumer, but sometimes fail. From Farmer's perspective, however, the bulk of global preventable disease is not only biomedical but also rooted in abuses of human rights. Indeed, for him, the ongoing death of children from preventable disease itself constitutes the greatest infraction of human rights. The DRC case should make it clear that this is not always to be understood as the result of a 'market failure'. Further, it is clear that oppression and exploitation can be key factors in the growth and sustainability of value chains, like the one that leads from a DRC mine to the smartphone. There is every sign that GAVI's activities can correct failures in the vaccines market to the direct benefit of children. I would not suggest that GAVI and the Foundation are ignorant or heedless of the links between disease and human rights abuses. But, in order to make a specific problem tractable, the frame they deploy does sideline those links. In doing so it tends to de-politicise the matter of childhood and disease. Philanthrocapitalists effectively control many of the resources that can be brought to bear on childhood disease. The question I would raise is whether it is therefore right for them also to control what can be said about childhood disease and how it should be framed.

Do vaccines create wealth?

As I suggested above, Dr. Barnado had to work hard raise funds. He did so, in part, by making use of the then culturally dominant notion of the salvation of the immortal human soul from Hell, however conceived. It is clearly a good thing to save a child's life by providing them with sufficient nutrition and so on, but Barnado offered more than this. He offered donors the opportunity, through charitable giving, more closely to emulate their deity and better to approximate a state of salvation. He also offered them the opportunity to rescue children from conditions that might incline them to sin, thus bringing their spiritual salvation into view. His philanthropy, then, was about far more than life or death because each individual child's existence was about more than their individual life or death. Children's health and well-being always matter for one reason or another and these reasons help to shape the steps that are taken to advance them.

Today's medical researchers are in a similar position to Barnado. When seeking charitable donations to fund research they have to persuade their sponsors that they can bridge the gap between wealth and good works. Like Barnado they have to bear in mind the kinds of culturally dominant ideas that their audiences might find persuasive. Farmer tells us a lot about poverty and oppression as factors in the distribution of disease. Other recent sources have also been at pains to highlight links between disease and the distribution of poverty but from a rather different angle. For Moxon et al. (2011) vaccines do not just save lives but also create wealth. Clearly just as social injustice reduces individuals' and communities' ability to resist infectious disease and to survive its effects, so the burden of health care and the loss of productivity consequent upon disease could be reduced by vaccines. Disease is always a bio-social as well as a biomedical matter. But it is worth paying close attention to what Moxon et al. (2011) are doing beyond simply stating facts. The paper is a call for increased participation and investment in GAVI programmes by nation states. Saving children's lives is presented as an investment in individual and national wealth. A healthy, well-developed populace will better be able to participate in the conversion of natural resources into internationally tradeable products and services. Just as in Barnado's day, children's lives are never simply a 'life or death' matter. Reasons are given for helping them to live that construe them as resources of one kind or another. For Barnado, spiritual resources, for Moxon et al. (2001), economic resources, and for Farmer (2004), resources for communities to resist oppression. The point here is not that it is wrong to consider children as resource, but rather to recognise that the question of what children are 'for' remains an open question and, given that answers to it are shaped by distinct and competing perspectives, a major political question.

Life and death and Bill Gates

Earlier in this chapter I raised the issue of good intentions and actual outcomes. I suggested that they needed to be understood in the context of the relationships formed between states, philanthropists and others actors. The work of GAVI and the Foundation on making vaccines more readily available to children of the majority world is a fine example of good intentions leading to benefits in practice. This work emerges from recently rearranged relationships between philanthropic foundations, donor and aid recipient nations and pharmaceutical corporations. These relationships have allowed the Foundation to leverage its donations into a broader form of influence that encourages the use of a philanthrocapitalist frame to understand the health and well-being of many populations of children. Just as GAVI changes market conditions, so the Foundation plays a part in shaping the conditions in which health and development are analysed, discussed and, so, rendered tractable. By contrasting a philanthrocapitalist frame with that used by Farmer (2004), I emphasised its tendency to treat issues that may be considered matters of intense political controversy as matters of technique.

It is not part of my agenda to assert that either Farmer's or a more philanthrocapitalist frame is the better. It would be possible to point out that the same sets of activities that have created a global elite have also wrecked the chances of many communities to support themselves through their own labour and, thus, to sustain their children's health. It would further be possible to fashion this connection into the idea that figures like Gates are tainted with hypocrisy. This, of course, would be to ignore the great deal of good contemporary philanthropy is currently doing and the good it may do in the future. Likewise, it would be possible to point out that Farmer's account of links between poverty and health as based in structural violence, that is itself generated by the greed and selfishness of relatively powerful individuals, is an account that is available within many global religions. It would further be possible to point out that these major global religious organisations have, at best, a patchy record of reducing that greed and selfishness. This sceptical view would tend to ignore the critical connections Farmer makes between the suffering of the many and the wealth of the few in this very unequal world.

There is a point, however, at which I would lodge special criticism of Gates' philanthrocapitalism in terms of what it implies for the multiplicity of voice. The following section is a transcript of a discussion broadcast in the BBC current affairs programme 'Newsnight' in 2011 as part of a longer examination of GAVI and philanthrocapitalism.

The first speaker is Deborah Done of the World Development Movement (www.wdm.org.uk). The second is Bill Gates.

> Done: 'Few would disagree with vaccinating millions of children ... the issue we have is that it's a bit of a distraction, so these sorts of top-down business-led philanthropic solutions distract from the bigger picture.... It's all well and good to be vaccinated against ... preventable diseases but if you send those families back out and they don't have land on which to farm because its been grabbed by big corporates ... or if they can't feed their families because of speculation on food prices ... because they are spending 90 of their income on everyday food stuffs so they can't educate their children ... all of these things will be masked and lost and we will be having the same discussion in 20 years time ... the reason those bigger issues aren't dealt with ... is that progressive taxation could be far more important, dealing with global monopolies could be far more important enabling people to have the policies to feed themselves would be a much more effective solution for dealing with poverty than vaccination programmes.'
>
> Gates: 'This is about the children and the mothers of those children and whether we take the technology that every rich child takes for granted and make it available ... if she has a scheme to change the economic world order that's all well and good, in the meantime, let's not let the millions of children die.'

The first observation I would make is that Done is attempting to bring a similar frame to Farmer's to bear on the issues. For her, high rates of preventable disease among the poor are a reflection of unjust economic relations. She refers quite clearly to the possibility of adjusting taxation, preventing monopolies and limiting corporate power to lay claim to land. For her, it seems, specific political decisions about the regulation or otherwise of global business have helped to create poverty. This means that the right policies successfully implemented could reduce poverty, bringing about a more just settlement and improving children's health.

Bill Gates' response is an interesting one. He responds to Done as if she had called for a change in the entire economic 'world order' rather than the adjustment of specific policies. He also focuses the issues on mothers and children. He further presents the issue in terms of the immediate life or death of millions of children. It is as if the life or death of children were a point of such simplicity that no debate is necessary. This chapter has detailed just how complex the relationships involved in GAVI's own efforts have been and how they have carefully tried to influence markets and governments. Gates' actions in fact address a similar mixture of powers to those listed by Done. However, when challenged in public debate, Gates talks as if the issues were simple.

There are many circumstances in which it would make very good sense for Gates to imagine himself in the position of a mother whose child is dying of a preventable disease. It generates sentiments of sympathy, some anger and frustration and a sense of urgency. Such an exercise of identification may well motivate his philanthropic efforts. However, in the context of this televised discussion, it tends to close down debate and to define alternatives as negligent. It places certain of his actions beyond question by marking the topic of the life and death of children as lying beyond the realm of political debate and controversy. We might take the view that since Gates' philanthropy shows every sign of success, he deserves the privilege of defining the scope of debate that he takes upon himself. What is lost in defining children's health exclusively as an emergency that must urgently be tackled is exactly the longer term perspective that Done insists on.

6
Childhood, Climate Change and Human Agency

The issues that arise at the conjunction of childhood and climate change are complex. They touch on the oldest themes of bio-politics; relations between human and non-human life; the nature of human agency and its traction and influence in the world; the possibilities for deliberate intervention in human and non-human activity. Over the past couple of centuries at least, children have had these themes projected onto their lives and childhood itself has been shaped by its use as a political and scientific test-bed for their examination. Looking to futures, these themes also have profound practical and ethical implications for children, for the distribution of resources essential to life and for intergenerational justice and generational succession (Facer 2011). So as I examine climate change and childhood together, I will advance an argument about the roles that children and childhood can play in responding to climate change. Central to this is my concern that children be seen not only as the resources out of which futures are built – so many bodies to be healthily grown, so many minds to be cultivated – but also as active and creative participants in the construction of futures. To this extent I'm clearly going to be building on views of children as agents that have been developed in recent decades in childhood studies.

My argument hinges on another, perhaps, less familiar view, however. The climate scientist Hulme (2009) is critical of what he calls the 'problem-solution frame' that is often applied to climate change phenomena. Following Hulme, I take the view that climate change is not best understood as a problem to be solved. I will suggest instead that changes in climate are better understood as set of phenomena that we humans now have no option but to participate in and, given their global scope, to inhabit. Climate is an open-ended, dynamic and changing phenomenon. It has no natural resting state and obeys no innate teleology. Human individuals, communities and societies certainly have preferences about climate and weather, but there is no universal consensus on these. Given this, if we find aspects of climate problematic, we should not suppose that we will be able to make it settle down into a pattern that all will find favourable.

There is further conceptual shift I am working with here. Some of the phenomena that comprise climate change might conventionally be described as 'social', while others might conventionally be described as 'natural'. There are, for example, specific rates at which different plant species under different conditions can absorb atmospheric CO_2; and, specific desires for convenience, achievement and economic success often inform individuals' transport choices. Crucially, however, climate change emerges from the dynamic ordering and reordering of relationships between and among such processes however we might seek to categorise them. In the terms I developed in Chapter 3, climate change phenomena are a set of 'biosocial events'. Because climate change is never only natural or only social it does not respect the distinctions that shape much extant biopolitical understanding. Consistent with arguments I've made in earlier chapters, when bios and zoë, lifestyles and life processes or the social and the natural are so tightly intertwined with such apparently dramatic consequences as global changes in climate, applying a frame that supposes them to be separate would be a questionable starting point (Latour 2004).

This presents a serious challenge to the analytic value of attempts to characterise climate change as a whole or in part as either a matter of scientific fact or as matter of human value, as either a matter for experts in the physical and life sciences or for social scientists. Since these phenomena cannot properly be categorised, it is difficult to subdue them through the application of appropriate expertise. It is this hybrid and unruly character of climate change, the way that it confounds many standard approaches to gaining traction on human conduct and the non-human world that leads me in this chapter and the next to rethink children as participants in the creation of futures.

If climate change cannot be 'solved', then the key question that faces present and future generations is how to participate in and to inhabit a changing climate well. Developing arguments that have already been made in climate change science circles (Hulme 2009), I will suggest that while the identification and solution of problems will remain a crucial skill and core cultural and technical practice in everyday life, and while greater levels of scientific literacy in populations are desirable, another practice that has cropped up elsewhere in the book and that I refer to here as 'reframing' will be essential in addressing that question of how to live well in relation to climate change. Currently a good deal of educational effort is made to familiarise children with modes of scientific practice and discovery and to apprise them of scientific facts. Far less effort is devoted to enabling children to define the challenges and opportunities that they see arising, to change the list of questions posed. If, as I suggested in Chapter 1, the binary frame remains a prominent feature of many scientifically informed cultures, yet it is losing its value as a means of gaining understanding and purchase on our circumstances, new generations are one place we may look for transformative insights. My suggestion, then, is that children are well placed to develop new ways to navigate the bio-social (Lee and Motzkau 2011). It is not my aim in what follows to answer the question of how to live well in the context of climate change, how to form preferences or how best to express them, but to suggest that children may be reconsidered as imagineers whose capacities to reframe problems and concerns need to be recognised, fostered and treated as one basis of action among others.

My preference for 'discussing responses to climate change phenomena' over 'trying to solve the problem of climate change' may be making some readers feel short-changed or even suspicious. After all, if we are not trying to solve what is often presented as one of the world's biggest problems, we might appear to be doing nothing at all. These feelings, if you have them, are getting right to the heart of the matter. The images of competence, expertise and ability to gain traction on the world that often crop up when distinctions are drawn between adults and children frequently hinge on the idea of successful deliberate and rational problem solving. When one sets the world in order according to one's intentions, one disavows the dependence, even helplessness, that is so strongly associated with infancy and childhood and asserts one's agency. This transcending disavowal is a key element, not only of some biographies and adult self-images, but also, as Latour (1993) has argued, of modern technological culture. He sees a bio-political commitment to the distinctiveness and autonomy of humans within the wider world as an implicit element of technical rationality. He also sees it as mistaken in so far as it ignores the fact that greater human technical capability means greater dependence on materials and non-human life-forms and their innate tendencies.

For my own part, even given a series of unintended consequences of attempts to solve problems – the use of chlorofluorocarbons in refrigeration to solve the problem of food storage, the use of fossil fuels to solve the problem of economic underdevelopment – this problem-solving, self-esteem building mindset is not necessarily a bad thing. That is, as long as it is applied in places where it has a chance of 'working', meaning places where its intended effects far outweigh its unintended effects. But as I have suggested, climate change does not respect the founding distinctions of human scientific expertise or images of human agency. If there are good reasons to see climate change as less than amenable to 'solution', then new ways of configuring human competence that are less reactive against feelings and states of dependence may be of value. In this context 'reframing' issues and concerns is a key but often uncelebrated human competence.

I have broken my coverage of climate change and childhood down into two chapters. This chapter describes the emergence of climate change into popular awareness and some aspects of its reception. It is concerned to establish that one major response to climate change has been to see it as a problem with a clear cause and a set of clear candidate solutions. This way of responding to climate change figures it as a problem that has had a linear development and that awaits linear solution. It retains the faith that solutions are available on the basis of a clear separation between social and life processes, between human action and outcomes. In the following chapter, I lay out different ways of characterising climate change that, in my view, better reflect our current circumstances. To reframe climate change as a 'wicked problem' (Brown et al. 2010) or as an 'emergent biosocial phenomenon' (Lee and Motzkau 2012b) or as an opportunity to engage 'the human spirit' (Hulme 2009) is to call for responses to it that are not about providing solutions to known problems but that rest instead on human capacities imaginatively to reframe their perceptions and assumptions. I will describe the operation of 'reframing', illustrate its role in addressing biosocial issues of the past and relate it both to children's education and to issues of intergenerational justice. I will argue that we need to change our minds about the kind of resource children are from human embodiments of the future to future makers.

Climate change and the 'Facts'

Contemporary discussion of climate change as a result of the use of fossil fuels and whether and how to respond to it is riven with controversy (Giddens 2009). There is a broad scientific and political consensus that climate change is taking place, that it poses major and multiple threats to human well-being and that action of some kind must be taken to limit its negative consequences (IPCC 2007). Some, however, argue against the idea that global climate is changing in any unusual or significant ways. Others accept that there is such change but reject the view that human lifestyles have much to do with it (Lawson 2009). Others still accept that anthropogenic, or human caused, climate change is taking place, but, on the basis of cost-benefit analysis, do not think that it should be treated as the most pressing challenge to human well-being (Lomborg 2001).

Many debates over whether or not anthropogenic climate change is taking place and whether or not urgent action is required proceed as if the phenomenon as a whole could be reduced to a point of scientific evidence. For example, the 'hockey stick' graph that showed a close relation between global atmospheric temperature and industrialisation (Gore 2006) has been used to present climate change as an 'inconvenient truth' lying outside and beyond human preferences and perceptions. The hope behind such a presentation is that if climate change can be established as a phenomenon that lies beyond or outside the realm of politics in which perceptions and preferences conflict, then concerted, unified human action will follow. On this view, 'facts', as determined by scientific investigation, can be relied upon to create consensus, and, further, consensus is the only basis for effective response.

For some commentators, however, it has become clear that assertions of the 'facts' have not led to consensus (Hulme 2009). For them, the politics of perceptions and preferences are best treated as an intrinsic part of the phenomenon of climate change. Certainly, perceptions of the value of future human life and preferences about the lifestyles that today's children can expect as adults do not derive or depend upon careful measurement but, nevertheless, remain central to our concerns. On this view, alongside the clashes of proponents and deniers, climate change also needs to be understood and addressed as a 'social fact' something that has real and significant consequences despite and, in some circumstances, because of doubts about its existence.

For these reasons, it is not my purpose in this chapter to assert an orthodox version of climate change. This is not because I doubt that climate scientists have privileged access to the data that accurately describe state of the climate, but because I am approaching climate change both as a social fact and as a set of phenomena that far exceeds any set of data points. I am paying attention to how perceptions and preferences are interwoven with factual accounts, especially where imaginations are brought to bear on devising suitable responses to present or projected challenges. So I am concerned with how climate change is construed, with how children figure in attempts to respond to it and how this interacts with their status as human fragments of the future.

For me, as well as being a social fact in the way I have described, climate change is also a space of governance, of practice and of biosocial imagination. This space has some interesting features that have implications for the distributions of Life, Voice and Resource that comprise childhood. First, many populations have come to aspire to a better life for their own children and to expect to see overall generational increases in health, wealth and longevity. Western post-industrial lifestyles are one of the templates for such expectations. The problem is that the abundance and choice of food and the inexpensive transport these rest upon are largely dependent on the use of fossil fuels. Thus, the phenomenon of climate change carries the implication that we need to reverse this arrow of expectation and accept the possibility that future generations will have less abundant lives if we are to forestall truly dire consequences. Second, if our efforts to limit or to adjust to climate change are to be successful, they need to happen on a very short timescale. As the time of writing, one campaign, striking a notably apocalyptic tone, has it that we have only 50 months to 'save the world' (www.onehundredmonths.org).

Children's status as human fragments of the future has long made it appear possible to bring about behavioural, economic and political change within the timeframe of a generation. On many estimates however, if we are to respond effectively to climate change, whether by reducing its rate or by making plans to deal with its consequences, we need either to act now or to have acted yesterday. Thus, climate change poses challenges for understandings of childhood and for the government of generational succession fostered within and intrinsic to Western modernities and related aspirations for economic and social development. For a few generations, the fact of youth has seemed to place the possibility of progress towards a better, more abundant world within practical reach on the timescale of about 20 years and children have been understood as making the problem of the future tractable as long as they learn their lessons and develop good habits. But a different frame is now arising in which present-day and future children figure as the inheritors of failure, and as the janitors of development, consigned to cleanup duties after the party is over.

In what follows I will sketch key understandings of how human lifestyle is interacting with life processes to precipitate climate change. I will consider the forebodings of finality and apocalypse that condition some response to climate change and draw attention to how this helps to figure climate change as a problem that ought to have a solution even in the face of its complexity.

Climate change as a complex challenge

Climate change has been described as the largest, the most threatening and the most complex challenge facing humanity (IPCC 2007). It is widely understood to involve the countless moments at which mundane aspects of human lifestyles that use fossil fuels for energy – such as transport, cooking, manufacturing and farming – interact with the content of the atmosphere through the emission of gases like carbon dioxide and methane. These gases are understood to increase the atmospheric 'greenhouse effect' that prevents the radiation of heat energy from the surface of the Earth into space. There are many life processes that remove CO_2 from the atmosphere. As plants grow, they convert CO_2 into climate neutral carbohydrates. Many microscopic sea organisms turn CO_2 into calcium carbonate, to build their exoskeletons. However, current use of fossil fuels like oil, coal, lignite and natural gas produces CO_2 at a pace that exceeds the capacity of those life processes that perform automatic 'carbon capture'. The net result is an overall warming of the atmosphere. CO_2 greenhouse effects are joined by those of methane gas (CH_4) that is, among other sources, a by-product of the digestive system of cattle and other livestock.

The topic of methane presents an opportunity to introduce the concepts of positive feedback and non-linear change and the part they play in climate change narratives and in the logic of climate fears. In the relatively low temperatures and high pressures that are found on ocean floors, a solid chemical compound known as 'methane clathrate' that is composed of water ice and methane can be as stable as water ice is in a functioning domestic freezer. Methane clathrate deposits effectively lock methane away from the atmosphere. It is hypothesised that rising sea temperatures will melt these clathrates, releasing vast quantities of methane into the atmosphere, thus accelerating global climate change (Kennett 2002). It is calculated that the methane is a much more potent a greenhouse gas than CO_2 (ibid).

This methane clathrate hypothesis is a good example of a form of positive feedback. As increasing concentrations of CO_2 bring about a temperature rise, this kicks off another process – melting methane clathrates – that feeds back into the existing change to accelerate warming. It is also a good example of a non-linear change. Even if atmospheric CO_2 concentrations increased in a linear fashion – by roughly the same amount every year – thanks to positive feedback from melting clathrates, the outcomes in terms of degrees of global warming and climate change could worsen very suddenly. Further, such a non-linear change might also prove to be irreversible. Even if, following a sudden spike in atmospheric methane concentrations, CO_2 concentrations were steadily reduced year on year, because of the new melting clathrate issue it is unlikely that a similar reduction in average global temperature would result. Positive feedback, non-linear change and irreversibility are features of systems in which multiple parallel processes are capable of interacting with one another. Further sites for positive feedback and irreversible change include the negative impact of increasing temperatures on those populations of plants and calcium carbonate producing animals that currently perform a good deal of carbon capture. Positive feedback, non-linear change and irreversibility are features of systems in which multiple parallel processes are capable of interacting with one another once their outputs reach given thresholds. There is no bar against 'social' processes of economic development and national and individual self-definition and 'natural' processes interacting one with another in this way.

These non-linear irreversible changes in biosocial relations are good examples of what I have described in Chapter 3 as biosocial events. Understanding them and finding practical responses to them require a facility that exceeds and any particular version of disciplinary expertise. This is because bio-social events do not respect the expectation of timing and interaction held either by life or social sciences (Lee and Motzkau 2011).

The aggregate effects of countless moments of contact between lifestyles and life processes are, on many scientific and governmental views, being registered today in an increase in the number and intensity of 'extreme weather events' such as droughts, floods and heat-waves. This has given rise to concerns that, across the current century, there will be massive decreases in the quality and sustainability of human life in many global regions, with shortages of food and water, and increases in rates of infectious disease and armed conflict (Dyer 2010). In the medium-term future, it seems, many people will suffer and die as a result of mundane aspects of human lifestyles in the present day. The speed at which climate change takes place may also threaten the viability of non-human species, plants, animals and bacteria, wherever they are unable to adapt quickly enough to new conditions through physiological or behavioural change, or migration. It seems urgent that something be done.

After two or three decades of international debate, the scale of the issue and the threats it poses are well rehearsed. The complexity of the challenge that climate change presents as an issue of governance is, however, harder to articulate. This is partly because it has taken decades of controversy and policy failure to reveal it. Climate change can be considered a complex challenge in at least three ways. It presents attempts to gain traction on it with paradoxes of agency and of desirability in the way of lifestyle and, deeply connected as it is with matters of value and of preference, it consistently overspills attempts to contain it within the bounds of expert knowledge.

The paradox of individual agency

Many of the mundane aspects of human lifestyles that contribute to, say, CO_2 emissions can be seen in terms of the actions of individuals. This makes the problem of climate change at once more and less tractable. From one perspective, the closer to our everyday lives the challenge lies, represented in the airmiles of our foods and the energy efficiency of domestic appliances and airplanes, the more easily it can be addressed by individual and collective agencies on a relatively small scale. The power to live more sustainably, it would seem, is in our hands. From a second perspective, however, the more mundane the problematic connections are, the more closely and inextricably they are woven into the pattern of everyday lives, and so the less open to change they are for the simple reason that so much of basic importance depends on them.

In recent years this contradiction has been eased somewhat from the point of view of the individual consumer by the development of new product ranges and services that enable them to make a personal choice for greater sustainability. But, where nuclear and renewable energy sources are not used, even electric cars ultimately draw their power from fossil fuels. Likewise, even if new airplane designs increase fuel efficiency, if more journeys are made, more fossil fuels, on aggregate, will be used. Resolution of this paradox would require either near complete transition to non-fossil fuel power sources or significant change in understandings of what is a desirable lifestyle, or both. This takes us to the next level of complexity.

Desirable lifestyles and economic growth

At the time of writing, countries of the European Union and the United States are experiencing severe economic turbulence. Such are the apparent dangers that climate and sustainability issues are being sidelined in political discussion. As a citizen of the United Kingdom I have become used to hearing news reports about national and regional rates of economic growth. When I hear that economic growth is slowing, this makes me worry for my future in retirement and for the lifestyle prospects of my baby boy. It is customary to read low growth figures as undesirable and a high growth rate, even though it brings its own tensions, as desirable. At the same time, however, it is clear that increased economic growth translates into increased rates of the conversion of natural resources into products. This has implications for climate change.

Over the past century or so, growth in the use of petrochemicals has leveraged growth in many other economic sectors. As fuels and fertilisers, for example, petrochemicals increase the rate at which farming can convert water, air and grain into foodstuffs. Likewise, air travel and container shipping have boosted global trade, increasing the rate at which resources are exchanged across geographical space along gradients of profitability. So, even as high rates of economic growth look like good news in the present and for some limited aspects of the future, such as my likely lifestyle prospects in retirement, they also look like bad news in the longer term for a wide but, as yet, unknown range of polities and people. Even today droughts and floods that have been connected to climate change (Sutton and Dong 2012) are increasing the cost of basic foodstuffs. At the time of writing, potato and other root vegetable prices have been affected in the United Kingdom as crops rot in sodden soil and, globally, wheat markets are being shaped by a Russian drought.

Economic growth rates are a key feature of the international governmental discourse through which the question of what is desirable in the way of lifestyle is currently decided. This discourse circulates among the many governments and international bodies such as the International Monetary Fund and Organisation for Economic Cooperation and Development that take on the task variously of enabling and imposing these lifestyles in anticipation of the desires and interests of their populations. In their favour, rates of economic growth provide a relatively durable basis for international comparisons that undergird many, arguably, progressive, initiatives. For example the pursuit of economic growth is a major motivation for improving children's access to education. On the other hand, as some now argue (Boyle and Simms 2009), as standardised metrics of comparison, measures of growth also sideline debates on alternative understandings of what is desirable in the way of human lifestyles. When economic growth is paired with climate change it seems that not only the means by which governments attempt to provide security for populations but also the standards by which their success is compared and judged are driving many populations into a paradox of desirable lifestyle. Fossil fuel–based economic growth will provide desirable lifestyles in the present for some while undermining the life chances of as yet unknown groups of people in the medium-term future. It is not at all clear that this paradox can be resolved. There does, however, appear to be a range of options.

Wherever there is doubt that life processes are abundant enough to sustain human lifestyles, can match the needs of a growing population or can survive the unintended consequences of our actions there are calls to limit 'growth' whether that be of human populations or of economic activity. In 1798, for example, Malthus (2008) predicted that the population of Britain would inevitably outstrip the ability of farmland to provide food. Unless population numbers were limited, mass starvation would result. The methods he used, however, took account neither of possible improvements in the management of soil fertility nor in the possibility of large scale international trade in food commodities. In 1973, the Club of Rome forecast, within decades, a major decline in the scale and quality of human life as a result of the exhaustion of natural resources (Meadows 1972). Happily this forecast has as yet also proved false. Lomborg (2001) points out that those forecasts took no account of innovation in the efficient use and recycling of resources. The idea of limiting human population as a way of resolving our paradox remains persuasive to some (www.populationmatters.org) as does the idea that economic activity could deliberately be reduced (Foster 2008). The deliberate imposition of such limits is less persuasive to key players in the regulation of the global economy. Christine Lagarde, the current director of the International Monetary Fund, for example, favours 'sustainable' growth and development – an emphasis on technological developments and business practices that are more efficient in their use of natural resources. There are also hints that she is persuaded of the need for reduced expectations of living standards in the majority world. For example, she took the drop in living standards experienced by many Greek citizens in the current Euro currency crisis as an opportunity to express her greater sympathy with the poor of sub-Saharan Africa.

It is clear then that just as fossil fuel use can be understood to drive change in climate systems, so data about them is driving change in social systems. Our paradoxes mark the thresholds at which such change could become non-linear and irreversible. If reductions in quality of life are at all controversial, then it should be clear, as I have suggested above, that 'climate change' is not just a matter of factual claims about natural systems, but that challenge of climate change now includes processes and forces of political controversy, behaviour and expectation. Climate change and a wider concern with sustainable living are now a resource for political and commercial decision making about the distribution of other resources. That is one of the things that sceptical commentators on climate change object to (Lawson 2009; Lomborg 2001). On Hulme's (2009) view, their objections can best be seen as another aspect of the complexity of climate change. This aspect directly affects the tractability of climate change just as much as the technical difficulties of powering our lives without fossil fuels do. Beyond this, a key question in the climate change debate is what we take a 'better' life for present and future generations to mean. It seems that as long as the aspirations of states, private concerns and individuals are captured by the goal of economic growth, this question will be relatively muted.

Technological solutions?

If my account so far bears scrutiny, it is clear that climate change is a pervasive, possibly irreversible threat not only to human well-being but also to many of the wider life processes that we rely upon. Further, it mixes political and scientific concerns in such a way as to make it hard to grasp within the terms of conventional distinctions between forms of expertise and ways of gaining traction on and influencing events. One result of this complexity is a series of paradoxes which appear to preclude effective response. It also appears to call for rapid responses. The apocalyptic tone of those who would 'save the world' in the next 50 months may not be fully merited. After all, it is only the world as we know it that is under threat. Even if climate change carries off most of the human population, 'life' will continue. But to say that the threat is largely to things that human value does not diminish it by much. Preferences to preserve our ways of life, our comfort, health and prosperity provide strong and understandable motivations to find a solution that can, somehow, cut through the complexities of climate change. Even if there is no clear way to navigate the biosocial complexity of climate change, couldn't the Gordian knot simply be cut in such a way as to preserve and perhaps extend existing wealthy lifestyles?

If we suppose that climate change is driven, at least in part, by increasing levels of CO_2 in the atmosphere and that this is, at least in part, a result of human activity, we might seek to change human activity in such a way as to reduce our CO_2 output. Given that there are tensions between reductions of CO_2 output and economic development and that different countries and global regions are in economic competition with each other, it is clear that a global solution will require international negotiation and agreement. But what happens when, as in Copenhagen in 2009 (Rogelj et al. 2010), these negotiations fail to produce an agreed outcome that is adequate to the task?

Alternative approaches that use technical means directly to target those atmospheric CO_2 levels certainly are open to us. These include the range of schemes and strategies for removing CO_2 from the atmosphere known as 'geo-engineering' (Goodell 2009) and the use of nuclear fission to generate electricity as an alternative to fossil fuels. One class of geo-engineering strategies that is under discussion (Boyd 2009) – artificial trees and scrubbing towers – would pass air through a filter made of sodium hydroxide which chemically reacts with CO_2 to form a solid compound which can then be buried, thus removing CO_2 from the air. This is a rather energy-intensive approach. The artificial trees and scrubbing towers would have to be built and the sodium hydroxide manufactured. One wonders where that energy is supposed to come from. The advantage of this approach is that if it is adopted on a large scale and for any reason its operation has unexpected and undesirable outcomes, the trees and towers can be deactivated.

A second approach to geo-engineering often referred to as 'ocean nourishment' would involve deliberately increasing the available levels of minerals, such as iron, in the upper levels of ocean regions where low nutrient levels currently limit the growth of phytoplankton (Dyer 2010). These organisms would then, in their greater numbers, and as an automatic feature of their life processes, go on to turn CO_2 into carbohydrates through the process of photosynthesis. One of the attractions of this approach is that once the iron has been produced and distributed in a suitably biologically available form, very little further fossil fuel–dependent activity would be necessary. The increased levels of iron would, in effect, enlist the capabilities of phytoplankton to use solar energy to capture CO_2. The strategy appears rather less attractive when we consider that once those extra nutrients are made available in fresh ocean regions, they cannot very easily be removed. If unexpected and undesirable consequences arose, there would be no obvious way to intervene.

Long a bugbear of environmental campaigners, nuclear power had been seeing a reversal in its fortunes over the past decade, gaining in popularity as a response to climate change with both campaigners and policy makers (Lovelock 2009). That was until a tsunami flooded a nuclear power station in Fukushima, Japan in 2011. The resulting meltdown of the facility killed several people, released radioactive materials in the surrounding area and led Japanese people to question the safety of food and water of Japanese origin.

If we suppose that technological developments that aim directly to reduce CO_2 can cut through the complexities of climate change to offer a clear solution, a clear navigable path through the new biosocial terrain, the examples above strongly suggest that we are mistaken. Just as climate change hybridises life and social processes, bringing their interaction and interdependence to light, so it challenges the vision of autonomous, technological problem-solving action that forms the base of many understandings of human agency. In the case of rising CO_2 levels, we see how closely unintended consequences can keep pace with and even overshadow the intended beneficial results of economic development. As we then considered possible solutions, whether based in individual behaviour, international negotiation or technologies, we saw a similar confounding of the deliberate and willed with the inadvertent and undesirable. One commentator (Dyer 2010: 21) expressed these issues in terms of mathematical complexity theory as follows;

> The essential insight in complexity theory would be: Don't think of this as a linear process. Think of it as a process where at any point, some small change of inputs could produce a massive, unexpected flip in outputs.... Expect that any solutions you apply are likely to further disturb the system, leading to an infinite series of surprises. Very different from the kind of approach that is often taken in public policy, which is that you only need to do THIS, and the problem will be solved now and forever.

Conclusion

Even though we may wish to see climate change as a problem to be solved, it clearly cannot be relied upon to confirm our view. If the assessment above is correct, then new understandings of human agency and of the ability to gain traction on the world are needed that do not suppose that we can solve the problem. In both childhood studies and in social science more generally, 'agency' can be understood either as an essence of humanity that is either enabled or suppressed by circumstances or as an emergent property of human interaction that has no settled character, that depending on circumstances may properly be considered the property of individuals or collectives, even of collections of human and non-human elements (Latour 1993). Much of the time, as in my earlier discussion of children's responses to the Mosquito teen deterrent, this is a distinction without a difference. The theoretical question of whether agency is being created or released by arrangements of media is not the central issue because ethical matters of justice take precedence. But in the case of climate change it would appear that the nature of agency is a primary issue. If an existing model of agency, oriented to problem solving, that is incorporated into the assumptions and basic practices of policy making is revealed as inappropriate or inadequate, we might find it helpful to think of agency as something that can be built and built in alternative ways. If the human agency we currently rely on, that is so committed to adult expertise, emerges from patterns of interaction, then building new forms of agency will mean putting people in different situations and rearranging relationships between and among them. In my view this will and should involve an expansion in the cultural repertoire of positions that are available for children as agents. It will also involve keeping the question of what is to count as a 'better' future for children an open one. In the following chapter I will expand on this line of reasoning to highlight some new ways of thinking about how children and human futures may be connected. We are used to thinking of children and to treating them as human fragments of a future that we can shape by shaping them. We are less accustomed to thinking of children and treating them as creators of futures precisely because their problem-solving agency is in question.

7
Children's Roles in Responses to Climate Change

In the previous chapter I introduced climate change phenomena and drew attention to the challenges they present to various performances of human agency. First, some individuals find *both* that their mundane daily activities contribute to CO_2 emission and thus present excellent sites for the application of their own initiative *and* that, precisely because these activities are mundane, they are so tightly woven into everyday life as to frequently place them beyond change that is practicable at an individual level. Second, policy makers who are committed to the security of their populations are aware of the possibility of insecurities arising from fossil fuel use, but are also aware that fossil fuels offer a reliable way of boosting the economic growth and development that, conventionally, make for security. Even where the IPCC view of climate change is accepted, these two paradoxes can hinder political attempts to intervene in climate change. A third challenge arises as a consequence of the complexity of the relations that comprise climate change – feedback, and non-linear and irreversible change. Deliberate technological interventions of sufficient scale significantly to reduce CO_2 levels may have unpredictable, irreversible and potentially negative outcomes.

On this view, instances of human agency taking the form of individual action, international political action and collective technological action would all seem significantly compromised. Unfortunately, this list includes major ways of performing agency that have found their way into the governmental common sense and 'best practice' of technologically developed democracies; individual action inspired by either self-interest or morality; the politically negotiated settlement of conflicting interests; attempts to make physical processes predictable and controllable. Even if my account is somewhat exaggerated in its pessimism, it is not at all clear that these conventional forms of agency, especially those that are designed for solving clearly defined problems, are fit for the purpose of responding to climate change. As a set of bio-social phenomena comprised from dynamic interactions between social and life processes, climate change is not a clearly defined problem. At the same time, large numbers of young citizens who, by virtue of their age, are the most likely to experience the results of present policy and activity are also, once again by virtue of their age and apparent immaturity, held outside many of the sites at which responses are formulated and considered.

In this chapter I will present children as an untapped resource for devising responses to climate change and the issues of sustainable living that it raises. I am certainly not alone in making a case like this. UNICEF, for example, has campaigns and programmes based on just that idea. UNICEF is especially concerned to promote children's engagement in majority world circumstances where adaptation to climate change is already a major aspect of everyday life. Thus there are practical and community-based initiatives that already rely on children's work and knowledge (Hart 1997). But I want to explore another area in which children may be central in building new forms of human agency, new ways to devise responses to climate change that might suit minority world circumstances.

As I have argued, climate change is testing the limits of currently dominant 'biosocial imaginations' – ways of thinking about human agency and about the traction it can have on social processes, life processes and their interrelations. I have also suggested that there is a danger that currently dominant intellectual and imaginative resources will fail this test. So I am going to consider what children might bring in the way of an intellectual contribution to addressing climate change. One of my motivations in this is ethical. I think it would be wrong if children were placed in the position of passively awaiting their role as the janitors of the future. But my thinking is also led by a reappraisal of the place of imagination and 'framing' within scientific discovery, technological application and the tractability of life and social processes that some commentators are now undertaking. Whether one is considering life processes (Wagner

2009), physical systems (Barad 2007) or political events and decision-making (Watzlawick et al. 1974), it is becoming clear that how one chooses to frame the issues under your consideration is of central and practical significance. I am going to consider the possibility, then, that children might help in 'reframing' climate change phenomena.

In considering this, I certainly will be met by the question of limits to children's competence. It might be said that even if they are best understood as 'agents' in relation to their family and community life, even in relation to a drought adaptation scheme in sub-Saharan Africa, that does not qualify them for involvement in responding to climate change in places like the United Kingdom. It can always be asked how fit children would be for the complexities of the bureaucratic engines of state, the drama of political leadership or the intensity of scientific research. These are legitimate questions in their own right, but they belong on a different axis to my concerns. I am not trying to establish a case that children have the same agency and the same competences that are often, and often complacently, attributed to adult individuals and collectives. Rather, in view of the limitations of these familiar agencies and competences, I will attempt to give expression to performances of agency and creative competences that, in my view, climate change phenomena demand, but that are relatively unsung and are very much open to the young. One way to make this agenda more explicit is by considering the various ways in which children are currently positioned in responses to climate change, so I will begin with a brief comparative survey of these.

How do children currently figure in 'climate response practices'?

In the last few decades, the rising profile of climate change has stimulated the development of a range of what I will call 'climate response practices'. These are activities that are undertaken in the hope and, often, belief that they will, in some way, affect climate change phenomena for the better. These practices vary widely from the establishment of 'transition towns' (Hopkins 2011) to citywide bicycle renting schemes and to public information campaigns that aim at changing individual behaviour (Whitmarsh et al. 2011). Here I will focus on just a few of those that involve or are conceived of as involving children directly. The establishment of shared and routine climate practices has been shaped and aided by existing forms of intervention and children have long been targeted as the recipients of scientific and civic education. Thus, the way climate practices frame children tends to overlap with a more general framing of children as relatively passive embodiments of the future.

1. Children, Scientific Literacy and Public Understanding of Science

Within this set of climate practices, children are understood as one of the many publics whose understanding of the content and status of scientific knowledge stands in need of improvement. This implies rather more than the simple transfer of facts from scientific communities to 'publics'. Improving public understanding of science is also an attempt to foster a critical and questioning approach to the many claims about processes, causes and effects that are advanced in popular media and political decision-making (Knight 2006). Where there is scientific consensus on a given matter, as with many climate change phenomena, it certainly is considered important within this practice that the content of the consensus is understood. However, the deeper understanding of science that is sought also involves an appreciation of the organised scepticism of the scientific community. The philosopher Karl Popper's (2002) views on the nature of scientific knowledge predominate here. On his view, hypotheses about processes, causes and effects can never be proven beyond doubt, though they can be rejected given the right evidence. Scientific knowledge, then, is distinguished by its inherent provisionality.

Building a critical awareness of the links between scientific enquiry, knowledge claims and policy means removing any expectation that scientific research provides a ground of certainty for policy making. It means accepting that all human knowledge is fallible, but also turning that fallibility into a tool for rejecting claims that have a poor or no evidential basis. The hope is that if publics can understand and accept this provisional relationship between science and authority, then passages between scientific research, democratic deliberation, and policy and practice will be far smoother than they are today.

Schooling activity and curricula tend to be closely regulated and schools are organised with the purpose of directing children's attention and efforts towards defined goals. Thus, where children are schooled, they offer a peculiarly available public for these practices to focus on. In UK schools at least these climate practices tend to have the relatively low ambition of delivering 'scientific literacy'. The current UK national curriculum content has been summarised as follows:

Age 5–11: Pupils should be taught to care for the environment as part of a topic on life processes and living things.

Age 11–14: Pupils should be taught how human activity and natural processes can lead to changes in the environment and about ways in which living things and the environment need to be protected. Teachers are encouraged to examine issues such as the finite resources available to us, waste reduction, recycling, renewable energy and environmental pollution.

Pupils demonstrate exceptional performance if they can 'describe and explain the importance of a wide range of applications and implications of science in familiar and unfamiliar contexts, such as addressing problems arising from global climate change'.

Age 14–16: Pupils should learn that the surface and the atmosphere of the earth have changed since the earth's origin, and are changing at present. They should also study how the effects of human activity on the environment can be assessed, using living and non-living indicators. Under 'applications and implications of science', pupils should be taught to 'consider how and why decisions about science and technology are made, including those that raise ethical issues, and about the social, economic and environmental effects of such decisions.'

(Guardian 2011)

I have suggested that a full public understanding of 'science' involves the simultaneous awareness *both* that there is a scientific consensus on many aspects of climate change *and* that all scientific knowledge, even a strong consensus, is provisional. On one view this would imply a contradiction – publics are asked both to accept and to suspend belief in scientific claims simultaneously. On another view, however, accepting and suspending belief and not being made too uncomfortable by that is a sign of kind of maturity, the kind that recognises that people, whoever they are, can only ever do their best to make sense of and respond within their circumstances. The apparent difficulty of achieving this familiarity with ambiguity and uncertainty underlies a good deal of the controversy and scepticism surrounding climate change (Norgaard 2011).

As the above quote indicates, although the UK curriculum certainly leads children to consideration of applications and implications of science, it does not require that schools and pupils deal with fallibility, arguably the key issue of the nature of scientific practice and authority and public relationships to it. There are many reasons why this issue might be downplayed. It may be that it is deemed to complex for young people. It may be that the fallibility of experts is considered a disheartening message. It may be felt to sail too close to climate change controversy. But the orientation to uncertainty that understandings of science appear to require is as emotionally based as it rationally based. It involves attitudes to authority, dependency and autonomy and to sources of comfort and hope. It involves changing any expectations that we might have of a general class of experts to sort out climate change on our behalf. Perhaps these emotional and attitudinal matters do not easily fit into a science curriculum. Perhaps this model of maturity in relation to scientific expertise sits awkwardly in those social contexts of education that depend, for good or ill, on distinctions of expertise and maturity between staff and pupils.

It is clear that this set of climate practices articulates children, schools (along with museums and scientific associations) and climate change phenomena, taking children as a particularly accessible public, and modulating its messages to suit school circumstances. The aim is to produce a future populace who are aware that climate change is an important issue, that there are means to produce reliable data about it and that ethical and policy issues are raised. As limited as they are, these objectives do not seem reasonable or appropriate to all commentators. Specific reference to climate change in science curricula in the United Kingdom is currently under review, for example (Department for Education 2011), and in the United States climate change teaching is becoming as controversial as other life science topics like evolution and creation science (Newton 2012).

2. Children as Future Scientists and Engineers

The previous set of climate response practices positions children as future members of a general population who will interact with science and technology primarily as citizens and as consumers. In contrast, this set of climate practices is concerned with children as future members of those segments of the population who are capable of developing and implementing technological responses and solutions. If the first was about developing understandings of expertise, the second is about developing experts. As I have suggested, climate practices tend to draw strength from and to be shaped by existing routines and ambitions. In this case, the desire to produce expert technological competence is oriented towards the ends both of economic growth and of developing responses to climate change.

The quality and quantity of science, technology engineering and mathematics (STEM) education in schools and universities is a matter of concern in many global regions. In Europe and the United States, deficits in the number of pupils who wish to study STEM subjects have been identified (Drew 2011) and have been woven into an influential policy narrative in which, unless the situation is improved, these regions are expected to lose out in competition with India and China. It is also widely understood that analytic and practical STEM skills will be of great value in devising responses to climate change, whether these be developing alternative power sources, using resources more efficiently or building resilience in the face of extreme weather events. There may appear to be a contradiction between participating successfully in a global economy-growing competition and responding to climate change. As I suggested in Chapter 8, this contradiction is often managed by reference to the 'green economy' and sustainable growth. If STEM skills are used to develop techniques and products that are profitable because they comprise effective responses to climate change phenomena, then it would appear that both ends can be served at once. This set of climate practices positions children clearly as relatively passive human fragments of the future. If economic and climate securities are to be available in that future, it appears vital that more children be educated in STEM skills.

3. Children as Passive Victims and Moral Witnesses

In the previous chapter in the context of my discussion of a paradox of individual agency, I pointed out that individuals are less free to alter their everyday climate relevant activities than they may appear, because these activities are woven into the fabric of their everyday lives. One way further to motivate adults to change their behaviour is to present their behaviour as comprising a threat to their children. There is a set of climate response practices that implicitly or explicitly uses this strategy. The idea that present-day and future children will be made to suffer as a result of present-day failures to 'save the world' is a widely available cultural trope. So widespread is it that when commenting, as I am, on childhood and climate change, it is actually quite difficult to avoid trading on it. Stern (2007) had to work hard to connect climate change with a threat to the economy. Arguably, it is easier to persuade people that there are threats to family, children and childhood.

A recent television advertisement, still available to watch on YouTube, commissioned by the UK Department for Energy and Climate Change and broadcast in 2009 is a good example of this set of climate practices. 'Bedtime Stories', part of the UK 'ACT ON CO2' campaign, shows a man (presumably a father) reading a young girl (presumably his daughter) a bedtime story about a land affected by extreme weather events (EWE). In an animation of the pages of the storybook, we are shown dry cracked earth, flooding, a crying toy bunny rabbit and a drowning puppy. The story continues to explain that scientists had discovered that the EWE were due to rising levels of atmospheric CO_2. The next plot point is the revelation that 40 per cent of CO_2 emissions arise in everyday electricity use for heating and lighting. The girl asks her father if the story has a happy ending. A voiceover informs us that it is in our power to affect the outcome of the story.

At the time of the broadcast the UK Advertising Standards Authority (ASA) received 357 complaints about the advertisement. Many wrote to contest the position taken that climate change was connected with human activity. Some argued that scientific opinion was divided on the issue. As with the emerging controversy about climate change and school curricula, this is a further illustration of just how complex the feedback processes are among climate change phenomena. On the basis of a set of findings and reports, a government cast climate change as a human made problem, partly rooted in everyday individual behaviour. It then commissioned an advertisement with the aim of influencing that behaviour. Aside from any intended effect it may have had, the advertisement stimulated a response that deployed the organised scepticism of climate research – the fact that disagreement is a normal, valuable feature of research communities – as grounds for the assertion that scientific opinion is divided and that, therefore, the £6 million pounds of tax-payers money invested in the campaign was illegitimately spent. Once again, unintended consequences would seem to be the rule rather than the exception when it comes to climate change phenomena.

Turning to the issue of children's positioning, the advert offers a sentimentalised view of the issues of intergenerational justice that climate change phenomena pose. It packages them in such a way as to stimulate a range of feelings – guilt, fear and a little hope. Rightly or wrongly, the assumption is that these feelings present adult viewers with powerful and salient incentives to change their behaviour. Whether or not this is accurate, it is clear that the advertisement positions children and childhood in specific ways. It plays into clichés about what children can and cannot understand, offering us a stereotypical 'child's view' of climate change, preferring animated toy animals to live action. In so doing it presents childhood as a state of innocence that is under threat even within the cosy confines of the bedroom. It presents a specific child as a witness to possibilities she can only just begin to comprehend, reliant on her father for information and comfort. The advert has her ask about the future of the imaginary land, while adult viewers know that that land is ours and hers. Aside from the knowledge the advertisement imparts, it also presents the girl as a witness to present day adult actions. My sense of this is that if anyone betrays that innocent-eyed witness, the hope of the future, they should feel ashamed. On this reading, the advertisement attempts to load the adult viewer with a sense of responsibility toward children and encourages the adult viewer to act on behalf of children.

4. Children leading climate change adaptation

The community of governmental and NGOs that concern themselves with 'development' in the 'developing' or 'majority' world face two problems that are of relevance here. They need to maximise the effectiveness of any programme they launch in the knowledge that the consequences of their actions can have direct positive or negative effects on the lives of vulnerable people, and in the knowledge that their budgets are limited. They also need to manage the legacy of imperial and colonial history in which minority world states and individuals assumed that they were equipped to make decisions on behalf of and to run the lives of majority world populations. These two needs, coupled with concerns for democratic accountability, and respect for human dignity, have made the participation of the intended beneficiaries a guiding principle of much development work (Hart 1997). Taking control of others' lives is expensive, can be counterproductive and is morally questionable from many perspectives. In contrast, approaches that enable people to make their own choices, applying their own preferences, values and knowledge are understood to foster synergies with the efforts of development agencies.

It is within this set of climate practices that children have been most clearly positioned as active participants in devising responses to climate change phenomena. Good examples of this positioning are offered by the Institute of Development Studies (http://www.ids.ac.uk). Extreme weather events, such as flood and drought, already affect children in the majority world and IPCC forecasts suggest that they will increase in frequency and severity in coming decades. When EWE meet poorly prepared infrastructure, decision-making processes and/or communities, disaster can result. IDS researches into and advocates the development of preventive and adaptive responses to such events that are 'child sensitive' in two ways; first, that they take children's specific vulnerabilities and needs into account; and, second, that they involve children's 'participation'. The version of participation IDS offer includes supporting children in their leadership of planning to prevent or adapt to EWE at the level of the home, school, community and beyond.

Commentary

Thus far, I have sketched just a few of the climate response practices that involve children directly. I have also drawn a contrast between those that position children primarily as relatively passive human futures and those that clearly envision children as having the capability, in the present, to make a positive contribution to domestic, community and regional responses.

It is important to note that these sets of practices are not mutually exclusive. It is entirely possible for the twin positions of 'human future' and 'present day agent' to exist alongside one another in documents, plans and activities. Indeed, as I have argued elsewhere (Lee

2001) this kind of overlap or ambiguity is a very widespread feature of contemporary childhoods. A good example of this overlap can be found in the US Environmental Protection Agency website (www.epa.gov/climate change). Here 'being part of the solution' involves learning about climate change, finding out how other people are responding to it and getting involved in mundane decision making about energy use.

I would suggest, however, that there appear to be two different scripts for children's involvement, one that is applied as a default in minority or developed world contexts and one that is available as alternative in majority or developing world contexts. It seems that where there are deemed to be adult experts – technical or political – in place who are already responding to climate change phenomena, children should be predominantly positioned as human futures, in receipt of preparation and training for their maturity. In contrast, it is only when adult expertise – technical or political – is deemed less available or, indeed, less than capable that children need be drawn on as resource in view of their extant creativity and knowledge. To the extent that climate change phenomena present problems that are being addressed and resolved by adult expertise, this arrangement of default and exception may well appear sensible. As I have suggested, however, it is not clear that these conditions hold.

I'd further note that where climate practices are educational in nature, the uncertainty of scientific discovery and the fact of abiding controversy tend to be marginalised. This may reflect pre-emptive responses to charges of the unwarranted politicisation of science curricula in view of the fact that even the IPCC consensus is not universally accepted. If so, this is rather ironic. As I have suggested, to say that climate scientists differ in their views and to use this as a reason to reject a given set of findings or a given consensus position is to misunderstand the status of scientific claims. Remember, Popperian orthodoxy has it that a hypothesis can never be proved, but only tested by attempts to find evidence against it. This organised scepticism and its implied difference of views is the normal condition of scientific research. If IPCC consensus is to shape policy and practice, that needs to be on the basis of a mature understanding of uncertainty and provisionality. If the experts sometimes get things right, that does not mean they should be followed slavishly. Likewise if they sometimes differ, or even err, that does not mean they should be ignored. Given this, it might not make much sense to shield children as learners or as participants from these difficulties. It may be better to involve them with and enable them emotionally and intellectually to deal with this uncertainty. It can be frightening to see that there are no positions of certainty and that leadership and expertise are fragile performances, but it can also be motivating.

So far, in this chapter, I have surveyed a few sets of climate response practices. My coverage has certainly not been exhaustive. There are, no doubt, many that I have not mentioned. From what I have seen, however, there seems to be a gap. In my view not enough is being done by states to develop active roles for children in devising responses to climate change phenomena in the present in technologically intensive, relatively politically stable regions such as Europe and the United States. Despite the work of such organisations as the UK Youth Climate Coalition (www.ukycc.org), the default assumption that children should be protected from climate phenomena by adults and by adult experts means that children in technologically intensive countries that have a relatively long history of stable democratic governance are insulated from the uncertainty and fallibility of scientific and technical knowledge and practice and have few opportunities actively to shape agendas.

I am not suggesting that it is always unreasonable for adults to shield children from threats and the knowledge of threats or from controversy and contestation. In my view, there are many circumstances in which this general protective attitude makes good sense. Rather, I am arguing that climate change phenomena present a rather special set of circumstances in which our default assumptions are a poor guide. First, if climates are changing rapidly, and if we can expect significant change on the scale of months and years rather than centuries, it might be pragmatically unwise to neglect children as sources of intellectual and practical resource. Second, if today's generation of children are going to have to cope with major changes in lifestyle across the course of their lives, ethical questions of intergenerational justice are especially acute. Third, if major extant forms of agency with respect to the bio-social – individual behaviour, political negotiation, technical fixes – are having difficulty gaining purchase on climate change phenomena, then old associations between adulthood and competence may no longer apply. These are the reasons behind my call for the development of new forms of agency, new ways of responding to climate change phenomena that better involve children and that do not require human action to transcend the non-human world. If climate change is not a problem to be solved, what other responses are possible and how do these relate to children?

Framing and reframing: Facts and values

We often think we can separate 'facts' and 'values' – that they can and should be teased apart. We might suppose, for example, that if only everyone could agree on the facts about climate change and leave their political viewpoints and vested interests aside, then the problem could be solved. It is also clear that individuals with different sets of values tend to prefer different sets of facts or interpretations of those facts. But something deeper is also going on. Where knowledge is produced in response to practical demands, facts and values have very intimate relationships and are held together and rendered consistent within shared 'frames'. One good example of a frame that holds facts and values together into an apparent consistency is the binary or two-category thinking that has long informed understandings of childhood and that, as I have argued, has kept the 'nature/nurture' question and its implied discontinuity between life processes of growth and social processes of socialisation alive in our culture even after its scientific value has waned. When states were trying to assess and to take hold of their populations as a set of resources, and to do so in way that could provide future securities, 'nature/nurture' was a useful frame for identifying both possibilities and limits to the power of interventions in children's lives to shape them in preferred ways. The 'nature/nurture' frame is not in itself 'factual'. Rather it expresses the values proper to states that have concern for their populations' security. It has, however, formed the basis for the production of factual claims in some areas of research. It provides the grounds, for example, for claims that comparative studies of the development of fraternal and identical twins can operationalise and answer otherwise nebulous questions about human existence, freedom and chance.

In taking this view of the power of 'framing' in generating and shaping bio-social knowledge, I am following Latour (2008) and

Stengers (2010). They argue that despite the dominance of a general view that facts and values are or should be separate, they are in fact intimately involved with each other. This does not lead these authors to the view that this involvement is simply an unpredictable mixture that cannot properly be analysed or studied. For them, finding such 'hybridity' is not a conclusion they come to but is their starting point. They argue that facts and values are mixed and related within 'matters of concern' and that this happens as people try to respond, for the best (however that is defined) within their circumstances.

How does this relate to agency?

One insight that I glean from Latour (2008), Stengers (2010) and the tradition of enquiry they represent is that we, human agents, need to recognise that we often see ourselves and base our understandings of what it is to be an agent as separate from and transcendent within a world of otherwise mute and meaningless matter. They argue that our situation and our options for agency are better understood on the basis of the view that we are in a condition of intimate connectability and separability (Lee 2005) with the materials that partly constitute our matters of concern. Our situation is such that the categories we use to understand our matters of concern and the expectations we form on this basis actually shape what we think it is possible for us to achieve and how we should go about planning our actions. The predominant frame we apply is the problem solving one that, in the previous chapter, appeared so limited in relation to climate change phenomena.

This connection between frames and agency has been well developed in the context of studies of psychotherapy and communication. Watzlawick et al. (1974), who is something of a sage in the varieties of systems theory that focus on feedback and irreversible change, has an excellent parable in this line that I paraphrase here:

> In a time of rebellion and political turmoil, the captain of an armed force faces an angry mob in a city square. His troops' rifles are aimed into the crowd. His superiors have commanded that he quell any riot using deadly force. He has the choice between firing, pretty indiscriminately, on a largely unarmed group of people, or

facing a firing squad himself on a charge of insubordination. What on earth can he do?

Thinking quickly, he climbs up high and makes it clear to the crowd that he is going to address them. As they listen, strongly motivated by the presence of rifles to hear what he has to say, he says:

'At the moment I can't tell which of you is a rioter and which of you are decent citizens passing through the square in the course of your legitimate business. I have strict orders to open fire on any rioters. Would those of you who are decent citizens please leave the square now so that my troops can fire freely?'

The square was empty within five minutes and the troops stood down.

In a context where a crowd of individuals had been identified as an obdurate problem that could best be solved by the technical means of shooting them dead, this captain followed his orders without killing anybody. In doing so he shifted between frames. In the frame provided by his orders, the discrimination between decent citizens and rioters had already been made by his superiors and was assumed to be a stable and obvious one. In the frame he mobilised, the issue of who was a decent citizen and who a rioter was turned into a question shared by him and the crowd. By reframing the situation, the captain expanded his options for action and created options for the crowd. Where problem-solving forms of agency are oriented towards finding the decisive action that will remove the need for further action, reframing is a kind of agency that can create more agents and forms of agency.

To develop this contrast further, problem solving tends to seek ways to transcend circumstances and to bring closure. In doing so it determines which persons or other resources can be part of the 'solution' from the outset. In contrast, reframing is oriented to widening the range of actions that are possible and towards communicating about preferences and values. In doing so, it sets new trains of thought and action going. For example, as Foucault points out in his discussion of changing responses to famine (see Chapter 2), where farmers desires for profit were once considered sinful and a cause of famine, as the issues became seen in economic terms, profit was revealed as an incentive for farmers to invest in their quality of their land. Farmers' interests became central to new more successful means of preventing famine.

It should be clear that it is not my intention to argue against a problem-solving orientation in general. There are many circumstances in which the questions of what is part of the problem and what is part of the solution have clear and stable answers. But these conditions are not always fulfilled. Indeed in the contexts of urban planning, sustainability and climate change where multiple interacting processes are at play, some authors have taken to using the term 'wicked problem' to describe situations that become more problematic and grow in their implications precisely as attempts are made to solve them (Brown et al. 2010). Wherever there is experience of erstwhile solutions transforming themselves into problems, modes of response that are based on reframing merit consideration.

Reframing the bio-social

It would be reasonable to point out that I have chosen to introduce reframing with a story that is exclusively about humans. Of course the Captain could reframe the situation. He could, after all, rely on his ability to communicate with the other human members of the crowd. But how could reframing be effective where significant elements of the situation are not human? One way to address this concern is to see it in terms of the zoë/bios distinction. It asserts the fundamental separateness of lifestyles that intrinsically involve communication and values, and life processes that do not. Agamben tells us that, despite its great significance and value, the zoë/bios distinction is itself a frame that was created to manage an ancient problem of government. To see life processes as intrinsically mute and meaningless and lifestyles as intrinsically communicative and value laden and, on that basis, to see the two as fundamentally separate is a perspective that may be chosen rather than a fact that must be obeyed.

A second response to the objection is offered by the evolutionary biologist Wagner (2009) in his work on relations between meaning, life processes and human choice. He points out that the life sciences are replete with paradoxes in which a given statement and its opposite are equally true. For example, it is equally viable to consider a cell as either a part of a larger organism or as a separate functional whole. It is the case that individual cells establish their own boundaries and regulate their own activities to a very large degree. It is also true that in doing so they rely on certain features of the organism they belong to or inhabit. It is clear that if the cell is removed from its familiar environment it may have difficulty maintaining itself and may die. But it is also true that a good number of these processes are concerned precisely to establish its difference from its environment, regulating, for example the passage of chemicals across its cell boundary. Wagner points out that which side of the paradox is preferred is never determined by facts, but is better understood as an analytic choice. He is clear that which choice is made has significant implications for the range of scientific questions that are asked and for opportunities to intervene in life processes. Indeed, for him, the human capability to frame and reframe life processes strengthens our ability to act on them, or in other words, deliberately to participate in them.

We can extend Wagner's view to the case of climate phenomena. See climate change as a set of independent problems and it makes sense to intervene by problem solving, ticking them off one by one. See it instead as a set of phenomena in complex and emergent relation and other responses may well be possible. It is likely that these will be organised not by the question 'how do we solve this?' but by the question 'how should we proceed, given these circumstances?' Rather like Latour and Stengers, Wagner sees a paradoxical condition of separability that we can navigate and work with as an intrinsic part of our ongoing matters of concern. Like them, he intends to raise awareness of the degree to which our frames shape our modes of response and of the degree to which we can change those frames.

A third response is simply to note that, as I argued in Chapter 8, and following Hulme (2009), the focus of human agency with respect to climate change phenomena is never simply on factual matters to do with life processes. This is because climate phenomena are biosocial in character, involving life and social processes together in such a way that making the zoë/bios distinction becomes less and less pertinent. Our situation is defined by the deep connectedness and mutual implication of lifestyles and life processes. When we meet climate change phenomena, we do not get away from values and preferences. Nor do we leave the possibilities of framing and reframing behind when we are involved in emergent bio-social phenomena. Indeed, there are reasons to think that reframing has been put to use with profound effects for many years.

In his account of bio-politics, Foucault frequently draws on instances where, over the course of decades, the basic frame that is applied to the matters of concern of states, rulers and others has been changed with dramatic results. In Chapter 2, for example, following Foucault, I described how famine and food supply came to be understood as matters of trade and economics rather than matters of morality and divine punishment. This had major, positive consequences for the security, health and well-being of some populations. Foucault does not offer us a general theory of how such changes occur. His methods were historical and empirical and he was closely attuned to contingency. One thing he certainly does not offer is an account of the changing of frames by an individual governmental genius. To a large extent, Foucault is interested in how frames change quietly in the background, as people try to cope with the matters of concern that present themselves. Sometimes frames change without anyone noticing. However, this does not exclude the possibility that as we become more aware of framing as a form of agency, more mindful of our own framing activity, we may choose to make deliberate use of it as a mode of response within the bio-social.

Just as Hulme (2009) identifies the limitations of 'problem-solution framing' of climate change, so he also explores an alternative frame. Rather than think of the climate as a set of problems to be solved he wants to see climate change used as a focus for the question of how to live well. For Hulme climate change phenomena may be novel, but they present humans with many fundamental issues that we are used to addressing without any expectation of solving them. They include issues of rights and responsibilities, relationships between generations, desires for personal growth and self-determination. As he puts it,

> ... we need to see how we can use the idea of climate change – the matrix of ecological functions, power relationships, cultural discourses and material flows that climate change reveals – to rethink

how we take forward our political, social, economic and personal projects over the decades to come.

(Hulme 2009: 362)

Bio-social education?

Reframing can be considered a form of agency. It works in the spaces between different versions of who or what is responsible for a given set of circumstances and between different versions of who or what has the power to direct change and stasis. It can also be considered a cognitive capability – we might call it 'practical imagination' (Lee and Motzkau 2012b). I need to be a little careful with the term 'imagination', especially in connection with children. It is all too easy to reduce it to wishful or unrealistic thinking – simply the opposite of the factual. In my terms, practical imagination is not like this at all. It is the human capability that remains effective when facts alone cannot be relied upon to guide our conduct. It is the faculty that enables us to become aware of the paradoxes and powerful choices that Wagner highlights. It brings a reflexive awareness of the frames we deploy. There is a risk that such awareness might lead to paralysis, to a reluctance to choose. After all, no guarantees can be made that the outcomes of the choice will be optimal. So practical imagination is a capability that involves courage alongside mindfulness. So what difference would greater attention to 'reframing' make for children and how would that engage them as active participants in climate practices?

If Popper's (2002) views hold water, then scientific practice is not directed towards the detection of certainties, but towards the systematic and strategic creation of uncertainties. For a number of reasons this is not always well understood. First, as the range and nature of uncertainties changes, so new aspects of the responsiveness of matter come to light and form the bases of new technologies. Since technologies are designed reliably to reproduce certain effects, they create localised and temporary pools of stability and capability for their users. The result is an end-user perspective on scientific practice that associates it with certainty. Second, some individuals and communities who are accustomed to receiving revealed truths from a deity who must not be challenged may take this frame with them as they participate in communication about life sciences. The result is a perspective from which scientific communications about evolution, climate change or genetic technology resemble false prophecy.

As I have suggested above, this tension between perspectives on science and certainty have left their mark on the way young people are educated about life processes. Science curricula often take on the difficult task of negotiating relationships to uncertainties and to authority on children's behalf. In doing so they take on a tension between highlighting the distinctive benefits of scientific practice while allowing that knowledge is provisional. Similarly, they try to avoid politicised clashes and present scientific practice as neutral and objective, while still promoting its centrality to human well-being. If we were to cultivate children's practical imaginations, then, formal education is clearly one of the structures we might try to use. A bio-social education that could usefully supplement current approaches, as I conceive it, would have four key features.

First, it would share a good deal of content and many topic areas with extant life science education, but it would strike a clear path away from concerns with the establishment of authoritative knowledge. Instead it would seek to generate the experience of paradox as Wagner describes it. To see that it is possible to view any living thing or life process – human or otherwise – both as an autonomous whole and as a part of a larger organised form may seem like a big conceptual leap for young people to make, but models of this abound in everyday social life. It is a standard aspect of children's socialisation in individualistic cultures, for example, to encourage them to act and respond sometimes as autonomous agents and sometimes as dependent and controllable parts of a family or a school. Knowing that the limits of personal responsibility and control are moveable and situational, that one simultaneously belongs to nobody and belongs to others, is a fundamental part of identity and ethical awareness (Lee 2005). Perhaps paradox is not so difficult after all.

Second, it would give children the opportunity to explore relations between fact and value not as they are ideally configured but as they are formed and reformed in practice. The idea that some knowledges are produced from a disinterested or objective point of view is often used to distinguish between trustworthy expertise and mere opinion. It most certainly is the case that scientific researchers often have special access to instruments and techniques for the production of data and are expected to make their methods transparent, so that others can check their working and results to see if they can reliably be reproduced. But that does not mean that individual scientists or research programmes are simply disinterested. Even as scientific research has a style of objectivity, it engages with matters of concern and clearly is an expression of human intellectual and moral values. Further, this is only one of the ways in which facts and values share the space of science. In many cases, scientific research is oriented towards the creation of profit and/or economic security. These different ways of living the relations between fact and value, these different matters of concern are part of different, but often overlapping networks of agents, that are themselves product of successive framings and reframings that have, over the years, established new relationships between states, businesses and knowledge. Finding ways to live relationships between facts and values involves social as well as intellectual skills. Bio-social education would give children practice in navigating circumstances where facts alone can no longer guide us.

Third, educational systems that divide up the universe of human understanding into sciences, arts and humanities implicitly make a statement to the effect that there are, and should be, distinct silos of human understanding. These silos have a long and complex history. But if bio-social education has the features I have previously described, it will in one way or another erode, confound or problematise them. Given the degree to which these silos are embedded in the institutional life of nations, adult biographies and some children's self-identities, it is important that this challenging characteristic of bio-social education is advanced strategically, choosing sites and opportunities with care.

Finally, bio-social education, as I envisage it, would tend to lower the age at which children can participate actively in the creation of responses to bio-social phenomena, including those associated with climate change. At each stage of bio-social education and across a range of topics, children will become used to asking a range of questions that Latour (2004), for one, would like to see embedded in adult political structures: Where and what are the relevant paradoxes? What styles of relating fact and value predominate and which are suppressed? What agents, processes and voices have been involved so far and which have been excluded? As they are equipped with these sensibilities, children will also be given opportunities to vault the barriers between stages of education. It would be a normal practice for children to spend some time working with their older peers.

Conclusion

Beginning with a presentation of climate change phenomena, I considered a range of attempts to address them as if climate change were a single problem that could be solved and as if the Earth stood in need of salvation through human action. As I did this I was concerned to highlight how modes of response contain images of and assumptions about human agency, its distribution across generations and its limits. To the extent that children are thought of as closer to 'nature' than adults, they are understood to be unable to play a part in responding to climate change phenomena. To the extent that adults are understood to be distant from 'nature' they are thought capable of the moment of transcendence of the non-human world that makes up problem solving.

My next step was to show how the 'problem-solution' frame has limitations that are exposed in practice when it meets the kind of 'wicked problems', full of unexpected and irreversible outcomes, that climate change entails. This led me to a pessimistic view about the future utility of the kinds of agency that are often, complacently, assumed to characterise adulthood. This paved the way for the consideration of other performances of agency, other modes of response that might prove at once more suitable for present circumstances and more congenial towards children's active participation in guiding responses. I found my model of alternative agency in the intellectual operation I called 'reframing' and then explored how it might help to shape 'biosocial education' that was designed to foster children's ability to respond to and transform existing relations among life processes and life styles.

The views I presented have an interesting relationship with the zoë/bios distinction that Agamben (1998) credits with foundational significance for bio-politics. Agamben is disturbed by the increasing tendency for the distinction between zoë and bios to be blurred in contemporary life and by contemporary forms of governance. In the 'states of exception' that he alerts us to, the sites at which government is made possible by the absence of the usual discrimination, he detects people being governed and managed on the basis of their life processes and in the absence of consideration of their abilities to reason and thus of their distinctively human value. In my terms he is warning against a collapse of the multiplicity of voice into the multiplicities of life and of resource. The line I have taken has a rather different take on the zoë/bios distinction. I have presented it as an obstacle to children's enfranchisement and as a hindrance to valuable modes of response to climate change phenomena. I have, however, remained sensitive to Agamben's concerns. I have tried to find ways to create desirable outcomes from that erosion of the zoë/bios distinction. It turns out that increases in children's voice and their deployment as creative human resources are at least thinkable in concert with greater awareness of life processes.

8
Conclusion

I started this book by describing what I called the 'survival fantasies' that surround children and childhood. Various survival fantasies depict a new generation or a new individual as a repository of hope, not only for the continuation of human life beyond that of present generations, but for the continuation of certain kinds, qualities and styles of life. To refer to them as 'fantasies' did not indicate that they should not be trusted, rather that they express values, desires and preferences. They communicate positions with respect to bios as well as zoë. Just as an individual child might be taken to embody physical features and personal characteristics common to their family, so a generation might find themselves expected to continue a culture or a lifestyle into the future.

Over the past few decades, some states have been quite successful in partially converting these desires into lived realities. The goal of making each new generation healthier and wealthier than their ancestors has become successfully incorporated into many state's functions as a key element of their temporal structure and commitments. This success has been particularly marked in recent years in states of the 'minority world'. Arguably, however, this form of progressive generational continuation has only been enabled by increasingly intensive use of 'natural' resources such as water, ore, soil, plant and animal fertility and antibiotic chemicals. This intensity of extraction, agriculture and manufacture has itself been enabled by the increasingly intense use of fossil fuels. It seems that we have now reached a point at which the unintended consequences of this development are eroding the possibilities of continuing in this way.

As it intersects with the human and social sciences, childhood research has long played its part in the development of states and in their ability to offer greater 'security', to their populations. When, as Foucault argued, questions of 'human nature' found a place in bio-political governance, children, often being seen as closer to nature, became the focus of the questions raised within the binary frame: How are nature and culture mixed within the growth of the child? Which characteristics are innate and which can successfully be encouraged or discouraged? For these reasons, childhood research came to be highly anthropocentric (Lee and Motzkau 2012b). This means not only that it was primarily interested in explaining and intervening in humanity, but also that it tended to disregard questions that lay beyond this agenda, beyond the binary frame. As long as the key bio-political questions of the day could be sketched as 'how can we civilize the young?' or 'how can we make the young fit to inherit our society?' this frame, for all its limitations, was a helpful navigational aid. But as new bio-political issues assert themselves, often under the heading of 'climate change', it is becoming clear that the space of 'human nature' is a relatively small one and that it could be unwise to consider it in isolation from wider bio-social concerns. Today's major bio-political questions seem to be more like 'how should we live if resources are not infinite?' and 'how should we relate to the life processes we depend on?' Each of these questions has implications for childhood and for generational succession.

A good deal of anthropocentric research has been concerned with the question of whether and how human individuals come to be separate from the rest of the world and thus transcendent over life processes (Lee 2005). Despite abiding critique from feminist (Barad 2007; Grosz 2004; Haraway 1991) and social scientific (Latour 1993) positions among many others, this view in combination with the 'technological sublime' (Nye 1994) has yielded a vision of human agency as control from a position of autonomy. This implies a self-sufficiency and separation that still informs hegemonic views of high-status adult agency. Others, writing from less anthropocentric perspectives and informed by their understandings of life processes, have sought to replace the question of how we get to be separate from and transcendent over life processes with insights into our abiding interdependencies on and alongside other life forms.

For the evolutionary biologist Margulis (Margulis and Sagan 1997), for example, human life is not just dependent on the plants and animals that we eat, but is equally dependent on less commonly regarded bacterial species. Among many other things, we rely on bacteria to decompose dead matter and so to make it biologically available to more complex organisms. If the bacteria involve in decomposition disappeared, we would soon be buried in waste. As discussed in Chapter 3, in terms of commensal bacteria, such dependencies exist not just in the environment external to the human body but in the environments that our individual cells encounter within our own bodies. Whether Lovelock's (2007) dire forecasts of climate change disaster are to be heeded, he has done much to frame human activity as just one, albeit growing, factor in the wider Earth system he calls Gaia. He sees humanity as inextricably linked with flows of matter and energy that link living things, rock, earth and sea together into a single system that can regulate its own temperature and atmospheric composition. This leads him to read future human deaths due to climate change as a form of Gaian homeostasis.

In my view responding to these changes in bio-political and intellectual contexts, in childhood studies as elsewhere, involves more than criticising a binary frame and more than rejecting previous research traditions. The response must also be creative. It means trying to find new ways of framing research questions, value positions and the desire to make a contribution to human well-being. In what follows, I will discuss some of the book's major concerns so as to highlight the opportunities for reframing that I have found.

Life, Resource, Voice

Moving out of the familiar and important but clearly limited anthropocentric space that has already been charted by the binary frame can provoke anxiety. Within the binary frame it has long been possible, for example, to know oneself as a 'progressive' or as a 'conservative' and thereby settle important questions of identity, group affiliation and moral purpose. The tendency has been for progressive commentators on human nature, for example, to invoke flexibilities of social life and human plasticities so as to imagine the techniques that could create a 'fairer' world, while conservatives would emphasise intractabilities of human desire and motivation. Bio-political issues tend to disturb these comforting identifications (Bull 2007).

Personal identity and moral purpose are of great importance in shaping research traditions, but the binary frame has also helped to guide estimations of what is and what is not a worthwhile research question (Lee and Motzkau 2011). Within its terms, for example, the age-related changes in high-frequency hearing discussed in Chapter 2 might be a near universal human developmental phenomenon, but they are marginal to the question of how the young, as human futures, can and should be shaped, since it's links with state organised practices of health promotion, education and training are, at best, trivial. To enter a space in which commensal bacteria, age-related hearing loss and global ecology are just as significant as child development, and in which the boundaries of the individual body and psyche are of interest only alongside phenomena that cross those boundaries, means entering a space that is only now being charted by childhood research.

That is why, in Chapter 2, taking cues from Foucault (2007) and Deleuze and Guattari (1998), I stepped to one side of the binary frame to find the abiding concerns that it addresses. The multiplicities I detected are emergent patterns of meaning, expectation and interpretation that, in my view, can be seen lending form to the discussions, strategies and interventions that compose childhoods. They do so when they are invoked by individuals and groups who are trying to respond to the practical problems and opportunities that face them. I do not advance these multiplicities as causal entities, rather as historically contingent regularities that themselves are open to change, but which, as an inherited cultural resource, also lend pattern to contemporary childhood bio-politics. Reliance on these multiplicities often goes unstated. They are often buried in the complexities of practice and policy and in the taken-for-granted manoeuvres that allow for practical navigation of these complexities. By bringing them to the fore, my intention is to open the operation of these multiplicities to critical scrutiny.

I have been able to illustrate the deployment and influence of these multiplicities as people try to figure out what they can and should make of children and childhood with examples of new deployments of old patterns and fresh life lent to old tropes. Examples of this include the way the assertion that it is natural to view human cognitive powers as a form of 'capital' folded a particular view of children as resource for the future into an account of life as 'flourishing', with the result that, in some circumstances, hope for the future can best be articulated only in terms of examination results (see Chapter 4). I also examined how the deployment of the mosquito teen deterrent bid to cut young people's voice and preferences out of the processes through which uses of urban space are negotiated in the hope of selecting between those aspects of youth that are a valuable resource to councils and shopkeepers and those that are not (see Chapter 2). Awareness of the same multiplicities helped to guide my examination of the complex links that are being established between the life processes of vaccine function, the use of children as resources in medical development and in the imagined futures of the majority world and the accountability of philanthrocapitalist activity (see Chapter 5).

There is no reason to suppose that multiplicities of life, voice and resource are the only way to navigate the bio-politics of childhood. Lupton (e.g. 2012), for example, has made excellent use of the critique of concepts of 'security' and 'risk' to challenge emergent taken-for-granted bio-political positions. Facer (2011), giving the emergent nature of futures and people's role in realising them their proper centrality, offers critical insights into understandings of how technological 'futures' can be made. Prout (2005) advocates strategic deployment of the 'excluded middle' as an alternative to the binary frame. But I think Life, Voice and Resource can be useful in comparing and linking areas of childhood bio-political activity that would otherwise appear to have little in common. In my view, they provide a script for the examination of childhood bio-politics that can lend the confidence needed to go to work on fresh areas. Considered as a navigational tool, then, these three multiplicities may assist in initial orientation to and path-finding within the world beyond the binary frame.

Bio-social events

In Chapter 3, I discussed a few approaches to childhood, individual development, behaviour and health that see humans in evolutionary context. These approaches, I argued, found it necessary to state their questions in terms of a number of overlapping timescales over which adaptations can take place. At the same time they considered individuals' and populations' health profiles and regularities in individual behaviour as the outcomes that emerge as these different temporalities come together in the present. In this sense, a number of life scientific approaches to childhood imagine form and conduct as 'untimely' (Grosz 2004), ruled neither by a static natural past nor by a changing social present. Within the binary frame it often makes sense to suppose that there is a conflict between different temporalities; on the one hand, if little is 'natural' a lot can be changed by the assertion of ethical and political positions; on the other hand, the quest for 'human nature' had such figures as Freud and Piaget enlist the child as a special piece of investigative equipment – closer to the deep natural past. Acknowledging untimeliness means reassessing the expectation that the temporalities of human life are stable and simple enough to build childhood research on.

For all their appreciation of the untimely, however, the more psychological of these approaches remain committed to the human individual as their unit of analysis. There is no reason why we should not be interested in questions of the formation of individuals, their conduct, functioning, feeling and thinking. But, as I illustrated in the case of the meningitis epidemic, bio-social events are not restricted to processes of individual development. Wherever a relationship between a life process of any kind and a lifestyle of any kind changes, we can speak of a 'biosocial event'. The concept can also be applied whether the event is brought about deliberately, as an unintended consequence of action, as the result of non-human life processes, or, as is often the case if it results from a complex intertwining of these. Thus bio-social events can be both a component of attempts to gain traction on life processes and lifestyles and a source of challenges. Seeing childhood and environmental challenges and opportunities in terms of the same analytic device arguably draws the two otherwise separate sets of issues into close alignment. I think this alignment is desirable in our present context where issues of sustainability, resource scarcity and intergenerational justice are raising questions of what can and should be made of childhood in response.

Children have long been considered human futures. Climate change is collapsing 'futures' closer into the present. In my view this calls for a response that can be delivered through relatively small changes in the nature of minority world childhoods. Rather than make individuals 'smarter' by feeding them cognitive enhancers, I ask how we can stimulate and foster children's practical imaginations to face up to the question of how to live well in the context of changing climate and resource scarcity. In Chapter 7 I advocated greater integration of children into scientific and technical controversy and ethical and political insight. I do not suppose that this will solve any problems, but I would expect it to help create the conditions for new forms of bio-social imagination that can reframe the apparently insoluble paradoxes of security we encounter today.

Bio-social imagination

My coverage of what I have referred to as the 'binary frame' has illustrated that the way we frame the contexts we find ourselves in, our own values and our agency has consequences for outcomes. One of the consequences of the binary frame in many states is a hegemonic network of associations between children, personal and national futures and expertise in development and socialisation. Another consequence is the presence of age limits to democratic participation. If Agamben (1998) is right, this 'inclusive exclusion' of children has been a key feature of bio-politics since Greek antiquity. The legitimacy of these arrangements rests, in part, on the fulfilment of promises of security and progress made by sovereign powers. I have argued that the possibility of imminent failure of these promises presents opportunities to reimagine and reframe relations between childhood, agency and expertise.

We have a degree of choice in how we frame our circumstances. Frames are expressions of value that are often informed by facts but are certainly not completed by them. Indeed, there is a range of authors including an evolutionary biologist (Wagner 2009), a climate scientist (Hulme 2009), a physicist (Barad 2007) who emphasise just how much 'choice' or opportunity for negotiation we have in the way we relate our understandings to our actions in the hope of producing desirable outcomes.

As I argued in Chapter 1, life processes are now centre stage in political thought and action. For Agamben this threatens a reduction in our status to 'bare life' – living in the exceptional spaces of sovereignty rather than in the dignity of the paradox of zoë and bios. I see it differently. There is an opportunity to see ourselves both as separate and as distinctive within a wider matrix of life processes – at once capable of autonomy, limited and enabled by contexts established by life processes, and thoroughly and inextricably embedded within and dependent upon these contexts. But we need to be mindful how we operate in this space so that we can devise appropriate performances of 'agency'.

Agency

One major theme in childhood studies over the past couple of decades has been the detection, theorisation and assertion of children's agency (Oswell 2012). The position I have developed clearly resonates with this. But it is also sensitive to a diversity of versions of agency, some of which have been identified as problematic. On the one hand, when we emphasise the separateness and autonomy of humans with respect to a wider matrix of life processes we reach for problem-solution responses to the challenges we meet. These rely for their effectiveness on the ability to escape unintended consequences of action. Today opportunities to escape these unintended consequences are growing scarce. On the other hand, when we emphasise our embeddedness and dependency and when our awareness of unintended consequences grows, we can lose sight of the tractability of the world and of the courage we need to adapt and to devise new responses.

For these reasons I was concerned in the final two chapters to imagine ways of fostering performances of agency that are far less influenced by the desire to escape dependency and feelings of weakness and insecurity. It is traditional in modern societies to see children as a way of making futures. I do not reject that tradition. So I spent some time thinking about how small changes to educational practices might help. You might have detected a note of complaint in Chapter 4 about the centrality of public examinations in education and their links with individual and international economic competition. In Chapter 7 I considered how educational practices might change if we saw outside the frame of 'schooling-work-economy'. I offered a few suggestions on how to foster young people's ability mindfully to reframe the problems they are presented with and courage to deal with uncertainty.

Outstanding issues

Much of this book has been taken up with the development and illustration of the range of conceptual tools and devices described above. It is important to note that they are not offered as theoretical hypotheses to be tested against empirical data. They simply do not have that structure or purpose. I do not suffer from the 'physics envy' that some social scientists are accused of and I do not believe that climate change and resource scarcity are the result of a deficit of data. If I were to compare reframing with another discipline or set of practices, it would be closer to 'engineering' than to any 'pure' science. For me, issues of possibility and tractability take precedence over issues of knowledge and authority. Having said this, the book also raises a range of important empirical questions that arise in the overlap between childhood and 'sustainability'. I'm going to bring to the book to a close by stating these as I see them.

Sustainability is a dominant metaphor in contemporary culture. It is used to frame problems and responses across a wide range of topics. Its resonance is due in part to the deep questions of human value and technical capability that it raises. Sustainability connects concerns about the opportunities and threats facing today's children and grandchildren with questions of how human lifestyles are, and should be, related to a wider matrix of life processes.

The term has long been applied to such uses of natural resources as packaging and energy. More recently, concerns for the supply and quality of food and water, the effectiveness of antibiotics and the economic practices of states, businesses and communities have been expressed in its terms. Thinking more broadly, extreme weather events call the sustainability of human habitation in some global regions into question. Despite the wide resonance of the term, however, a sustainable future is far from guaranteed. A shared concern across these topics is that today's unsustainable lifestyles will leave a bitter inheritance for future generations. In this context, I ask whether and how we can shape today's childhoods to create liveable and desirable futures. This enquiry has three main aspects:

1. Sustainable Education

   - What dispositions, aptitudes and skills will children need?
   - Sustainability crosses disciplinary boundaries. What are the implications for school and university curricula?
   - How can cognitive skills of systemic thought and 'reframing' be taught?

- How can curricula acknowledge scientific uncertainty and controversy?
- How can these issues be reconciled with skills for business, with qualification systems and with other trends in education policy?

2. Intergenerational Justice

- Can children and young people's participation in political and technical planning and decision-making be increased?
- What do they have to contribute?
- What practices could foster new forms of solidarity among the young and across generations?
- What practices might exacerbate existing inequalities and tensions of generation, class, ethnicity and gender?
- How will state and international governance preserve intergenerational contracts?

3. Lifestyles and life processes

- Are there generational differences in the way bio-social relations between human lifestyles and wider life processes are understood?
- Can youth be a test bed for new bio-social imaginations?
- How are young people's views of human/nonhuman relations shaped?
- What roles do faith and science play in this?
- How are values, responsibilities and techniques to be related in such areas as GM food and renewable energy?

References

Agamben, G. (1998) *Homo Sacer: Sovereign Power and Bare Life*. Stanford: Stanford University Press.

Alderson, P. (2000) *Young Children's Rights: Exploring Beliefs, Principles and Practice*. London: Jessica Kingsley.

Als, H. (1995) The Preterm Infant: A Model for the Study of Fetal Brain Expectation. In J-P. Lecanuet, W.P. Fifer, N.A. Krasnegor and W.P. Smotherman (eds) *Fetal Development : A Psychobiological Perspective* (pp. 439–471). Hillsdale, NJ: Erlbaum.

Annas, G. (2009) 'Global Clinical Trials and Informed Consent'. *New England Journal of Medicine*. 360, 20: 2050–2053.

Ariès, P. (1962) *Centuries of Childhood*. London: Jonathan Cape.

Barad, K. (2007) *Meeting the Universe Halfway: Quantum Physics and the Entanglement of Matter and Meaning*. Durham: Duke University Press.

Barnados (2013) http://www.barnardos.org.uk/barnardo_s_christian_heritage. pdf accessed 4 January 2013.

Barnett, M. and Weiss, T. (2008) *Humanitarianism in Question: Politics, Power Ethics*. New York: Cornell University Press.

Basu, K. (1999) 'Child Labour: Cause, Consequences and Cure with Remarks on International Labour Standards'. *Journal of Economic Literature*. XXXVII: 1083–1119.

Bishop, M. and Green, M. (2010) *Philanthrocapitalism: How Giving can Save the World*. London: A and C Black.

Bjorklund, D.F. and Pellegrini, A.D. (2000) 'Child Development and Evolutionary Psychology'. *Child Development*. 71, 6: 1687–1708.

Blakemore, S. and Frith, U. (2005) *The Learning Brain: Lessons for Education*. Oxford: Wiley Blackwell.

Bostrom, N. and Sandberg, A. (2009) 'Cognitive Enhancement: Methods, Ethics and Regulatory Challenges.' *Science and Engineering Ethics*. 15, 3: 311–341.

Boyd, P.W. (2009) 'Geopolitics of Geo-engineering'. *Nature Geoscience*. 2: 872.

Boyle, D. and Simms, A. (eds) (2009) *The New Economics: A Bigger Picture*. London: Routledge.

Brown, V.A., Harris, J.A. and Russell, J.Y. (2010) *Tackling Wicked Problems*. London: Routledge.

Bull, M. (2007) 'Vectors of the Biopolitical'. *New Left Review*. 45: 7–25.

Burman, E. (2007) *Developments: Child, Image, Nation*. London: Routledge.

Buroker, N.E., Ning, X.H., Zhou, Z.N., Li K., Cen W.J., Wu X.F., Ge M., Fan L.P., Zhu W.Z., Portman M.A., Chen S.H (2010) 'Genetic Associations with Mountain Sickness in Han and Tibetan Residents at the Qinghai-Tibetan Plateau'. *Clinica Chimica Acta: International Journal of Clinical Chemistry*. 411, 19–20: 1466–1473.

Burr, V. (1995) *An Introduction to Social Constructionism.* London: Routledge.

Buss, D.M. (1995) 'Evolutionary Psychology: A New Paradigm for Psychological Science'. *Psychological Inquiry.* 6, 1: 1–30.

Cipriani, D. (2009) *Children's Rights and the Minimum Age of Criminal Responsibility: A Global Perspective.* Farnham: Ashgate Publishers.

Cockman, P., Dawson, L. and Mather, R. (2011) 'Improving MMR Vaccination Rates: Herd Immunity is a Realistic Goal'. *British Medical Journal.* 343: d5703.

Cooper, C.L., Goswani, U. and Sahakian, B.J. (2009) *Mental Capital and Wellbeing.* Oxford: Wiley-Blackwell.

Coppens, P., Fernandes da Silva, M. and Pettman, S. (2006) 'European Regulations on Nutraceuticals, Dietary Supplements and Functional Foods'. *Toxicology.* 221, 1: 59–74.

Corsaro, W.A. (2004) *The Sociology of Childhood.* Newbury Park: Pine Forge Press.

Coveney, C., Gabe, J. and Williams, S. (2011) 'The Sociology of Cognitive Enhancement: Medicalization and Beyond'. *Health Sociology Review.* 20, 4: 381–393.

Crawford, D.H. (2007) *Deadly Companions: How Microbes Shaped Our History.* Oxford: Oxford University Press.

Credit-Suisse (2011) *Global Wealth Report* https://infocus.credit-suisse.com/data/_product_documents/_shop/323525/2011_global_wealth_report.pdf.

Darwin, C. (2003) *On the Origin of Species.* London: Wildside Press.

Dawkins, R. (2006) *The Blind Watchmaker.* London: Penguin.

DCDC (2007) The DCDC Global Strategic Trends Programme 2007–2036. http://www.dcdc-strategictrends.org.uk.

De Grey, A. and Rae, M. (2008) *Ending Ageing.* New York: St. Martins Press.

Deleuze, G. (1991) *Bergsonism.* Cambridge Mass: MIT Press.

Deleuze, G. and Guattari, F. (1998) *A Thousand Plateaus.* London: Athlone.

Department for Education (2011) http://www.education.gov.uk/inthenews/inthenews/a0073149/national-curriculum-review-launched. Accessed 6 March 2013.

DiMento, J.F.C. and Doughman, P. (eds) (2007) *Climate Change: What it Means for us, our Children and our Grandchildren.* Boston: MIT Press.

Donzelot, J. (1979) *The Policing of Families.* Baltimore: Johns Hopkins University Press.

Drew, D.E. (2011) *STEM the Tide: Reforming Science, Technology Engineering and Math Education in America.* Baltimore: Johns Hopkins University Press.

Durham County Council (2008) http://www.durham.gov.uk/durhamcc/pressrel.nsf/Web+Releases/9B151A656B3FD9AB802574CF002D51F1?OpenDocument. Accessed 31 July 2010.

Dyer, G. (2010) *Climate Wars: The Fight for Survival as the World Overheats.* Oxford: One World.

Elias, N. (1994) *The Civilizing Process.* Oxford: Blackwell.

Facer, K. (2011) *Learning Futures: Education, Technology and Social Change.* London: Routledge.

Farmer, P. (2004) *Pathologies of Power.* Berkeley: University of California Press.

Finlayson, C. (2009) *The Humans Who Went Extinct*. Oxford: Oxford University Press.

Forbes (2012) http://www.forbes.com/forbes-400/list/ Accessed 4 January 2013.

Foster, J. (2008) *The Sustainability Mirage*. London: Earthscan.

Foucault, M. (2007) *Security, Territory, Population*. New York: Picador.

Foucault, M. (2008) *The Birth of Biopolitics*. New York: Picador.

Freeland (2011) http://www.theatlantic.com/magazine/archive/2011/01/the-rise-of-the-new-global-elite/8343/ Accessed 20 October 2011.

Furedi, F. (2002) *Paranoid Parenting: Why Ignoring the Experts may be Best for Your Child*. Chicago: Chicago Review Press.

Giddens, A. (1992) *The Transformation of Intimacy*. Cambridge: Polity.

Giddens, A. (2009) *Politics of Climate Change*. Cambridge: Polity.

Gill, A. (1998) *Orphans of the Empire: The Shocking Story of Child Migration to Australia*. New York: Random.

Glenn, J.C. and Gordon, T.J. (2007) *2007 State of the Future*. New York: World Federation of United Nations Associations.

Gluckmann, P. and Hansen. M. (2006) *Mismatch: The Lifestyle Diseases Timebomb*. Oxford: Oxford University Press.

Greely, H., Sahakian, B., Harris, J., Kessler, R.C., Gazzaniga, M., Campbell, P. and Farah, M.J. (2008) 'Towards Responsible Use of Cognitive Enhancing Drugs by the Healthy'. *Nature*. 456: 702–705.

Grosz, E. (2004) *The Nick of Time: Politics, Evolution and the Untimely*. Durham: Duke University Press.

Goldacre, B. (2009) *Bad Science*. London: Harper Collins.

Goodell, J. (2009) *How to Cool the Planet*. New York: Mariner Books.

Gore, A. (2006) *An Inconvenient Truth*. London: Bloomsbury.

Gould, S.J. (2000) *Wonderful Life: Burgess Shale and the Nature of History*. New York: Vintage.

Government Office for Science (2008) *Foresight Mental Capital and Wellbeing Project: Final Project Report – Executive Summary*. London. HMSO.

Guardian (2011) http://www.guardian.co.uk/education/2011/jun/12/climate-change-curriculum-government-adviser. Accessed 30 November 2012.

Haraway, D. (1991) *Simians, Cyborgs and Women: The Reinvention of Nature*. London: Free Association Books.

Harris, J. (2007) *Enhancing Evolution: The Ethical Case for Making Better People*. Princeton: Princeton University Press.

Hart, R. (1997) *Children's Participation: The Theory and Practice of Involving Young Citizens in Community Development and Environmental Care*. London: Routledge.

Hope, J. (2008) *Biobazaar: The Open Source Revolution and Biotechnology*. Harvard: Harvard University Press.

Hopkins, R. (2011) *The Transition Handbook*. London: Transition Books.

Hulme, M. (2009) *Why We Disagree About Climate Change: Understanding Controversy, Inaction and Opportunity*. Cambridge: Cambridge University Press.

Hunt, L. (2008) *Inventing Human Rights: A History*. New York: W.W. Norton and Co.

Hutchby, I. and Moran-Ellis, J. (eds) (1998) *Children and Social Competence. Arenas of Action*. London: Falmer.

IPCC (Intergovernmental Panel on Climate Change) (2007) *Climate Change 2007 – The Physical Science Basis*. Cambridge: Cambridge University Press.

Jackson, T. (2011) *Prosperity Without Growth: Economics for a Finite Planet*. London: Routledge.

James, A. and Prout, A. (eds) (1997) *Constructing and Reconstructing Childhood: Contemporary Issues in the Sociological Study of Childhood*. 2nd ed. London: Falmer.

Jenks, C. (1996) *Childhood*. London: Routledge.

Jones, R.A.L (2007) *Soft Machines: Nanotechnology and Life*. Oxford: Oxford University Press.

Kellert, S.R. and Wilson, E.O. (1995) *The Biophilia Hypothesis*. Washington D.C.: Island Press.

Kennett, J.P. (2002) *Methane Hydrates in Quaternary Climate Change: The Clathrate Gun Hypothesis*. Washington: American Geophysical Union.

Knight, D. (2006) *Public Understanding of Science: A History of Communicating Scientific Ideas*. Abingdon: Routledge.

Latour, B. (1993) *We Have Never Been Modern*. Cambridge: Harvard University Press.

Latour, B. (2004) *Politics of Nature*. Harvard: Harvard University Press.

Latour, B. (2007) *Reassembling the Social: An Introduction to Actor Network Theory*. Oxford: Oxford University Press.

Latour, B. (2008) *What is the Style of Matters of Concern?* Amsterdam: Van Gorcum Press.

Lawson, N. (2009) *An Appeal to Reason: A Cool Look at Global Warming*. London: Gerald Duckworth and Co.

Lee, N.M. (2001) *Childhood and Society: Growing up in an Age of Uncertainty*. Maidenhead: Open University Press.

Lee, N.M. (2005) *Childhood and Human Value: Development, Separation and Separability*. Maidenhead: Open University Press.

Lee, N.M. and Motzkau, J. (2011) 'Navigating the Bio-Politics of Childhood: How Far can Hybridity Take us?' *Childhood: An International Journal of Child Research*. 18, 1: 7–19.

Lee, N.M. and Motzkau, J. (2012a) 'The Biosocial Event: Responding to Innovation in the Life Sciences'. *Sociology*. 46, 3: 426–441.

Lee, N.M. and Motzkau, J. (2012b) 'Varieties of Biosocial Imagination: Reframing Responses to Climate Change and Antibiotic Resistance'. *Science, Technology and Human Values*. July 16, 2012: 0162243912451498.

Lewin, R. (1994) *Complexity: Life at the Edge of Chaos*. Upper Saddle River: Prentice Hall.

Lock, M. and Kaufert, P. (2001) 'Menopause, Local Biologies and Cultures of Ageing'. *American Journal of Human Biology*. 13, 4: 494–504.

Lomborg, B. (2001) *The Skeptical Environmentalist*. Cambridge: Cambridge University Press.

Lovelock, J. (1979) *Gaia: A New Look at Life on Earth*. Oxford: Oxford University Press.

Lovelock, J. (2007) *The Revenge of Gaia*. London: Penguin.

Lovelock, J. (2009) *The Vanishing Face of Gaia: A Final Warning*. London: Allen Lane.

Lupton, D. (2012) *Fat*. London: Routledge.

Mahon, R. and MacBride, S. (2009) *The OECD and Transnational Governance*. Vancouver: University of British Columbia Press.

Malthus, T. (2008) *An Essay on the Principles of Population*. Oxford: Oxford University Press.

Margulis, L. and Sagan, D. (1997) *Microcosmos: Four Billion Years of Microbial Evolution*. Berkeley: University of California Press.

Meadows, D.H. (1972) *Limits to Growth*. New York: Signet.

Meadows, D.H., Randers, J. and Meadows, P.L. (2004) *Limits to Growth: The 30 year update*. New York: Chelsea Green.

Moore, S. (1992) 'Meningococcal Meningitis in Sub-Saharan Africa: A Model for the Epidemic Process'. *Clinical Infectious Diseases*. 14, 2: 515–525.

Mortensen, G. (2010) *Stones into Schools*. London: Penguin.

Motzkau, J.F. (2010) 'Speaking up Against Justice: Credibility, Suggestibility and Children's Memory on Trial', in J. Haaken and P. Reavey (eds) *Memory Matters: Contexts for Understanding Sexual Abuse Recollections*. London: Routledge.

Moxon, E.R., Das, P., Greenwood, B., Heymann, D.L., Horton, R., Levine, O.S., Plotkin, S. and Nossal, G. (2011) 'A Call to Action for the Next Decade of Vaccines'. *The Lancet*. 328: 298–302.

Murray, L. (1999) *The Quality of Sprawl*. Sydney: Duffy and Snellgrove.

Newton, S. (2012) http://www.earthmagazine.org/article/voices-defending-science-link-between-creationism-and-climate-change. Accessed 6 March 2012.

Ngai, P. (2005) *Made in China: Women Factory Workers in a Global Marketplace*. Durham NC: Duke University Press.

Norgaard, K.M. (2011) *Living in Denial: Climate Change, Emotions and Everyday Life*. Cambridge Mass: MIT Press.

Nye, D.E. (1994) *American Technological Sublime*. Cambridge Mass: MIT Press.

Office of Science and Technology (2005) *Drugs Futures 2025? Executive Summary and Overview*. London: HMSO.

Oswell, D. (2012) *The Agency of Children: From Family to Global Human Rights*. Cambridge: Cambridge University Press.

Panno, J. (2006) *Stem Cell Research: Medical Applications and Ethical Controversy*. New York: Checkmark Books.

Pantley, E. (2007) *The No-Cry Discipline Solution*. New York: Adaptations.

Parsons, T. (1956) 'The American Family: Its Relation to Personality and the Social Structure', in T. Parsons and R.F. Bales (eds) *Family, Socialisation and Interaction Process*. London: Routledge and Kegan Paul.

Piaget, J. (1927) *The Child's Conception of the World.* London: Routledge and Kegan Paul.

Popper, K. (2002) *Conjectures and Refutations: The Growth of Scientific Knowledge.* London: Routledge.

Prout, A. (ed.) (1999) *Body, Childhood and Society.* Basingstoke: Palgrave Macmillan.

Prout, A. (2005) *The Future of Childhood.* London: Routledge.

Qvortrup, J. (1994) 'Childhood Matters: An Introduction', in J. Qvortrup, M. Bardy, G. Sgritta and H. Wintersberger (eds) *Childhood Matters: Social Theory, Practice and Politics.* Aldershot: Avebury.

Rapley, J. (2004) 'Development Studies and the Post-development Critique'. *Progress in Development Studies.* 4, 4: 350–354.

Richardson, R.C. (2009) *Evolutionary Psychology as Maladaptive Psychology.* Cambridge Mass: MIT Press.

Rogelj, J., Nabel, J., Chen, C., Hare, W., Markmann, K., Meinshausen, M., Schaeffer, M., Macey, K. and Hohne, N. (2010) 'Copenhagen Accord Pledges are Paltry'. *Nature.* 464: 1126–1128.

Rose, N. (2007) *The Politics of Life Itself: Bio-Medicine, Power and Subjectivity in the 21st Century.* Princeton: Princeton University Press.

Sandel, M. (2004) 'The Case Against Perfection'. *Atlantic Monthly.* 293, 3: 51–62.

Schaffner, W., Harrison, L., Sheldon, K., Miller, E., Orenstein, W., Peter, G. and Rosenstein, N. (eds) (2004) *The Changing Epidemiology of Meningococcal Disease among US Children, Adolescents and Young Adults.* Bethesda: National Foundation for Infectious Diseases.

Seligman, A. (2009) *The Spoilt Generation.* London: Piatkus Books.

Sen, A. (2001) *Development as Freedom.* Oxford: Oxford Paperbacks.

Serres, M. (1995) *The Natural Contract.* Ann Arbor: University of Michigan Press.

Simopoulos, A.P. (2002) 'The Importance of the Ratio of Omega-6/Omega-3 Essential Fatty Acids'. *Biomedicine and Pharmacotherapy.* 56, 8: 365–379.

Smith, R. (2010) *A Universal Child?* Basingstoke: Palgrave Macmillan.

Stainton Rogers, R. and Stainton Rogers, W. (1992) *Stories of Childhood.* London: Harvester Wheatsheaf.

Stengers, I. (2010) (trans. Bononno, R.) *Cosmopolitics I.* Minneapolis:. University of Minnesota Press.

Stern (2007*) The Economics of Climate Change: The Stern Review.* Cambridge: Cambridge University Press.

Stewart, I. and Cohen, J. (1997) *Flights of Reality: The Evolution of the Curious Mind.* Cambridge: Cambridge University Press.

Sutton, R.T. and Dong, B. (2012) 'Atlantic Ocean Influence on a Shift in European Climate in the 1990's'. *Nature Geoscience.* 5: 788–792.

Sydenham, E., Dangour, A.D. and Lim, W.S. (2012) *Omega 3 Fatty Acid for the Prevention of Cognitive Decline and Dementia (Review).* The Cochrane Library, 2012 Issue 6. Oxford: Wiley.

US Department of Energy (2009): http://www.ornl.gov/sci/techresources/Human_Genome/home.shtm.

Vygotsky, L.S. (1986) *Mind in Society: Development of Higher Psychological Processes*. Boston, MA: Harvard University Press.

Wagner, A. (2009) *Paradoxical Life: Meaning, Matter and the Power of Human Choice*. New Haven: Yale University Press.

Wakefield, A.J., Murch, S.H., Anthony, A., Linnell, J., Casson, D.M., Malik, M., Berelowitz, M., Dhillon, A.P., Thomson, M.A., Harvey, P., Valentine, A., Davies, S.E. and Walker-Smith, J.A. (1998) 'Ileal-Lymphoid-nodular Hyperplasia, Non-Specific Colitis and Pervasive Developmental Disorder in Children'. *The Lancet*. 351: 637–641.

Watzlawick, P., Weakland, J. and Fisch, R. (1974) *Change: Principles of Problem Formation and Problem Resolution*. New York: W.W Norton and Company.

Weber, M. (2010) *The Protestant Ethic and the Spirit of Capitalism*. Eastford CT: Martino.

Whitmarsh, L., O'Neill, S. and Lorenzoni, I. (2011) *Engaging the Public with Climate Change: Behaviour Change and Communication*. London: Earthscan.

Williams, S. and Martin, P. (2009) 'Risks and Benefits may Turn out to be Finely Balanced'. *Nature*. 457: 532.

Wright, A. and Hastie, N. (2007) *Genes and Common Diseases: Genetics in Modern Medicine*. Cambridge: Cambridge University Press.

Zajda, J. (ed.) (2005) *International Handbook on Globalization, Education and Policy Research: Global Pedagogies and Policies*. New York: Springer.

Index